SpringerBriefs in Applied Sciences and Technology

Multiphase Flow

Series editors

Lixin Cheng, Portsmouth, UK
Dieter Mewes, Hannover, Germany

For further volumes:
http://www.springer.com/series/11897

Mingyan Liu · Zongding Hu

Nonlinear Analysis and Prediction of Time Series in Multiphase Reactors

 Springer

Mingyan Liu
School of Chemical Engineering
    and Technology
Tianjin University
Tianjin
People's Republic of China

and

State Key Laboratory of Chemical
    Engineering
Tianjin
People's Republic of China

Zongding Hu
School of Chemical Engineering
    and Technology
Tianjin University
Tianjin
People's Republic of China

ISSN 2191-530X   ISSN 2191-5318 (electronic)
ISBN 978-3-319-04192-6   ISBN 978-3-319-04193-3   (eBook)
DOI 10.1007/978-3-319-04193-3
Springer Cham Heidelberg New York Dordrecht London

Library of Congress Control Number: 2013956739

Printed on acid-free paper

Springer is part of Springer Science+Business Media (www.springer.com)

Acknowledgments

The authors are grateful to the Natural Science Foundation of China (contract Nos. 20106012 and 20576091), the Open Foundation of State Key Laboratory of Multiphase Complex System, Institute of Process Engineering, Chinese Academy of Sciences and the Cheung Kong Scholar Program for Innovative Teams of the Ministry of Education (contact No. IRT0641) for financial support of the work. The authors also wish to thank Dr. Klaren D. G., Profs. Kwauk M., Li J., and Liu C. for their helpful comments and suggestions. We want to thank the members of our research group team for their great contributions for this work.

Contents

Abstract

This chapter reports our recent work on the nonlinear aspects or deterministic chaos issues in the systems of multiphase reactors. The reactors include gas–liquid bubble columns, gas–liquid–solid fluidized beds, and gas–liquid–solid magnetized fluidized beds. Pressure fluctuations in the multiphase systems were taken as time series for nonlinear analysis, modeling, and forecasting. The qualitative and quantitative nonlinear analysis tools include attractor phase plane plot, correlation dimension, Kolmogorov entropy and largest Lyapunov exponent calculations, and local nonlinear short-term prediction. First, chaotic bubbling dynamics and transition mechanism in gas–liquid bubble columns with single orifice were investigated. Second, the chaotic characteristics in gas–liquid bubble columns with multiorifice distributor were presented. Finally, nonlinear behaviors in other types of multiphase reactors were explored briefly. The main results are as follows. Bubbling process with single orifice exhibited deterministic chaotic behavior in a certain range of gas flow rate. With the increase of gas flow rate, the sequence of periodic bubbling, primary chaotic bubbling, advanced chaotic bubbling, and jetting was successively found. The sudden increase of local nonlinear forecasting error of pressure signal with time provided an evidence of the presence of chaotic bubbling. Multivalue phenomena of chaotic invariants resulted from the multiscale behavior existing in the heterogeneous flow regime in multiorifice bubble columns. Flow regimes and their transitions could be characterized effectively by the chaotic invariants for these multiphase reactors. For gas–liquid–solid fluidized beds, some new flow regimes were identified by chaotic analyses. For gas–liquid–solid magnetized fluidized bed, the values of chaotic invariants reduced with the increase of magnetic field strength at given superficial gas velocity and liquid velocity.

Keywords Gas–liquid bubble column · Gas–liquid–solid fluidized bed · Time series · Chaotic analysis · Pressure fluctuation · Prediction

Notation

B	Magnetic induction, T
$C_{(r)}$	Correlation integral
d	Embedding dimension
D_2	Correlation dimension
f	Frequency, Hz
g	Functional relationship between the current state and the future state
G	Predictor which approximates function g
$H(x)$	Heaviside function
K	Kolmogorov entropy, bit/s
$K_{2,m}$	Order-2 Renyi entropy, bit/s
$L(t_k)$	Replacement length at evolution time t_k of k step, dimensionless
m	Embedding dimension
M	Total known number of data points
n	Total number of replacement steps
N	Length of restructured vector X_i
N_p	Number of predicted points
P	Pressure, p_a
P_e	Experimentally measured pressure, p_a
P_p	Predicted pressure, p_a
$P_{(f)}$	Power spectrum
Q	Gas flow rate, ml/min
r	Radius of hyper-sphere, dimensionless
X_i	Restructured vector
x	Component of restructured vector
t	Time, s
U_g	Superficial gas velocity, m/s

Greek Letters

δ	Standard deviation or size of the attractor
ε_g	Holdup, dimensionless
γ	Correlation coefficient
λ_1	Largest Lyapunov exponent
τ	Time delay, s
Δt	Sample time interval, s

Chapter 1
Introduction

Understanding the hydrodynamics in the gas–liquid bubble columns with single- or multi-orifice or in the gas–liquid–solid fluidized beds including gas–liquid–solid magnetized fluidized beds is of great importance for process industries but it is still not well known due to the limitations of traditional analysis methods and the complexity of multiphase flow systems themselves [1, 2]. It is interesting to study the bubbling hydrodynamics from the point of view of modern non-linear dynamics [3–10].

Regarding as the bubbling hydrodynamics with single orifice, Tritton and Egdell [11] carried out a set of experiments in a beaker on bubbling from an upwardly pointing vertical tube of an internal diameter of 1.2 mm. A period-doubling sequence leading to chaotic behavior in the inter-pulse return maps was found by analyzing the signal from a hot-film anemometer probe placed in the liquid near the bubble formation region. Mittoni et al. [12] performed a set of experiments over a slightly broader set of conditions than that of Tritton and Egdell [11]. Similar sequence was observed in the inter-pulse return maps by analyzing the gas inlet pressure signal in the gas chamber. Nguyen et al. [13] investigated the spatio-temporal dynamics in a train of rising bubbles and found that the analogy between the rising bubbles in fluids and an inverted dripping faucet is weak. Li et al. [14] studied the in-line bubble coalescence in non-Newtonian fluids in a bubble column. Bubble generation was through an orifice of varying diameter, submerged in the non-Newtonian liquid at the center of the bottom section of the tank. The signals of frequency of bubble passage at different heights in the tank were measured by optical probes of photodiodes placed at the external wall of the square duct. The nonlinear analyses revealed that the coalescence between bubbles obeys a chaotic and deterministic mechanism. Ruzicka et al. [15] studied the chaotic features of bubbling from a submerged orifice above a gas chamber using the simultaneous measurements of gas chamber pressure and liquid velocity fluctuations. Besides return maps analysis, mean and variance, intermittency factor, distribution of the length of jetting bursts, Kolmogorov entropy, K, correlation dimension, D_2, the Mann-Whitney statistic, power spectrum of signal were calculated to reveal the underlying structure of the data. Although the typical period doubling sequence leading to chaotic behavior was not found, the sequence of periodic bubbling,

M. Liu and Z. Hu, *Nonlinear Analysis and Prediction of Time Series in Multiphase Reactors*, SpringerBriefs on Multiphase Flow, DOI: 10.1007/978-3-319-04193-3_1, © The Author(s) 2014

chaotic bubbling and random bubbling was observed. Luewisutthichat et al. [16] investigated the power spectrum, K and D_2 by analyzing the rise velocity and shape fluctuations of single bubbles bubbling through an orifice in a two-dimensional bubble column and found that the signals of bubble velocity and shape exhibit chaos. Tufaile and Sartorelli [17] studied the dynamics of the formation of air bubbles in a submerged orifice in a water/glycerin solution inside a cylindrical tube. The delay time between successive bubbles was measured with a laser-photodiode system. The sequence of period-1, period-2, period-3, period-4, period-2 and chaotic bubbling was observed by analyzing the constructed bifurcation diagrams and first return maps. Tufaile and Sartorelli [18] studied the formation of air bubbles in a submerged nozzle in a water/glycerol solution inside a cylindrical tube submitted to a sound wave perturbation. They observed a route to chaos via period doubling as a function of the sound wave amplitude and established relations to a Henon-like dynamics with the construction of symbolic planes. They suggested that the bubble formation be seen as an oscillator driven by a sound wave. They [19] also found a route to chaos via quasi-periodicity.

Mosdorf and Shoji [20] described nonlinear features and analytical results for the chaotic bubbling from a submerged orifice of 2 mm in diameter and microconvection induced by the bubble generation was recorded using hot-probe anemometer located close to the orifice. They performed the nonlinear analysis for the time series data of hot-probe anemometer including the calculation of the largest Lyapunov exponent and proposed a simple model to simulate the process of interaction between the elastic bubble wall and liquid. They found that one of the reasons for chaos appearance is the nonlinear character of interaction between an elastic bubble wall and the liquid stream.

Sarnobat et al. [21] studied the effect of an applied electric potential on the dynamics of gas bubble formation from a single nozzle in glycerol experimentally. They found that at constant gas flow, bubble-formation exhibited a classic period-doubling route to chaos with increasing potential. Although electric potential and gas flow appear to have similar effects on the period-doubling bifurcation process for this system, the relative impact of electrostatic forces is smaller. However, the relative impact of electrostatic forces for the case of insulating liquid and conducting gas phases is comparable to flow forces.

Li et al. [22] studied the dynamics of a chain of bubbles rising in polymeric solutions by stimulus-response simulation, birefringence and particle image velocimetry. They identified two aspects as central to interactions and coalescence: stress creation by bubble passages and their relaxation. The competition displays complex nonlinear dynamics and can be described by a simple but explicit mathematical model. Recently, Frank and Li [23] investigated the chaotic dynamics of a bubble chain rising in a polymeric fluid by cognitive modeling and analytical analysis. Good agreement of the cognitive simulation results with experimental data was obtained and this reveals that the period doubling bifurcation sequence is the route to chaotic behavior.

Cieslinski and Mosdorf [24] discussed the nonlinear features of the air bubbling from a submerged glass nozzles submerged in distilled water in a cylindrical tank.

A laser-photodiode and an acoustic technique system were used. They found that bubbles dynamics is of deterministic chaos nature and behaviors of such system can be chaotic or periodic depending on the volume flow rate. Data recorded by acoustic system was used to analysis of bubble wall movement and found that the waves on bubble surface can be modeled by low dimensional model.

Garstecki et al. [25] studied rich nonlinear dynamics of bubbling in a microfluidic flow-focusing device. They observed period doubling and halving bifurcations and chaotic bubbling–characteristic features of a model nonlinear system. Their experimental observations indicate a dynamic similarity to a system that forms the topologically inverted version of the bubble generator –a dripping faucet.

Mosdorf and Wyszkowski [26] investigated the deterministic chaos appearance in bubbling flow. In the experiment, bubbles were generated from the brass nozzle with the inner diameter of 1.1 mm submerged in the glass tank (400 × 400 × 700 mm) filled with distillated water. Pressure fluctuations and signal from the laser-phototransistor sensor were recorded simultaneously. The movement of bubble wall was measured using a high speed camera and image processing technique. Two ranges of the air volume flow rate with different kinds of bubble chaotic behaviors were identified. For the air volume flow rate less than 0.2 l/min, pressure fluctuations are chaotic, but bubbles depart almost periodically. The reason for increase of chaotic character of the air pressure fluctuations is connected with increase of amplitude of pressure jumping after the bubble departure. For the air volume flow rate higher than 0.2 l/min, the departing bubbles join or group close to the nozzle outlet. This behavior of bubbles causes the increase of complexity of pressure fluctuations. The time of stability loss is approximately equal to average time between subsequent departing bubbles. When q increases over the 0.4 l/min, the rapid increase of complexity of bubble wall movement is observed—the system becomes almost stochastic.

Ruzicka et al. [27] studied experimentally bubble formation at two orifices with a common plenum by means of analysis of gas pressure fluctuations measured in the plenum. Two synchronous regimes were found at low and high gas flow rates, separated by a wide range of asynchronous regimes. Parameters like orifice spacing, water height, and column diameter influence the stability of sync regime.

As for the hydrodynamics in a bubble column with multi-orifice gas distributor, it is of great importance to identify the flow regime for proper design, scale-up, optimal operation and effective control of such reactors. But, a simple, objective, accurate and robust method for flow regime identification has not been well established and a combined use of many analytical tools is recommended in industrial practice. Meanwhile, new and effective analytical techniques for flow regime classification should also be developed. The flow regime identification is usually carried out by using some analysis techniques that characterize relevant invariant properties of the hydrodynamics. The analysis techniques used most often are as follows: visual observation and image analysis; statistic analysis; spectral analysis; Wigner analysis; Hurst analysis; wavelet analysis; chaos analysis, etc. Among these analytical methods, chaos analysis is considered to be one of the most potential methods. The most important reason is that chaos analysis can unravel more useful multi-dimension information about the nonlinear hydro-dynamics by processing a representative one-dimensional signal measured from the systems [3, 28].

Letzel et al. [28] characterized the regimes and regime transitions in bubble columns by chaos analysis of pressure signals. They found that the transition from the homogeneous to the heterogeneous flow regime in bubble columns can be quantitatively found with high accuracy by analyzing the chaotic characteristics of the pressure fluctuation signal. The Kolmogorov entropy as a function of gas velocity indicates a sharp transition from the homogeneous to the churn-turbulent flow regime and from other methods considered (e.g. holdup and other properties of the signal such as variance), this transition was less clear. Therefore chaos analysis is a powerful technique for on-line identification of flow regimes.

Femat et al. [29] carried out a set of experiments on bubbling in a vertical column to study its dynamical behavior. Fluctuations in the volume fraction of the fluid were indirectly measured as a time series and were used to reconstruct the underlying attractor. The characterization of a reconstruct attractor was carried out via Lyapunov exponents, a Poincare map and spectral analysis. They found that the bubble stream interactions induce the presence of complex oscillatory phenomena. For instance, since the bubbles rise almost linearly into the plume, the presence of the central plume induces an almost-periodic behavior. The number of fundamental frequencies (which are in some sense induced by the modes of the bubble streams) increases when the superficial gas velocity increases, yielding a route to chaos (periodic-quasiperiodic-chaotic behavior).

Chen et al. [30] analyzed the time series of local heat transfer rates measured by using a hot-wire probe in three bubble columns of different diameters of 200, 400 and 800 mm by means of rescaled range (R/S) and deterministic chaos analyses and found that due to the influence of highly chaotic bubble motions, the instantaneous local heat transfer exhibits low-dimensional chaotic features. The dependences of Hurst exponents and Kolmogorov entropy on the column scale consistently suggest different nonlinear hydrodynamic behaviors exist in bubble columns of different scales. An artificial neural network was applied to correlate instantaneous local heat transfer with dynamic motions of bubble and liquid.

In ordinary chaos analysis, the flow regime is also basically characterized and quantified by the chaos invariants, and the regime transition is identified when an abrupt or obvious change in a chaotic invariant is found. The chaotic invariant has to be estimated in order to study its tendency of change with the variation of the operation condition. However, the chaos invariant has multi-value feature in certain range of operating conditions. That is to say, more than one value of the invariant can be found under the same operating conditions [31–35]. For the multi-value phenomenon of the chaos characteristics, two different treatment strategies can be seen in the previous papers. One is to give only one value of the chaos parameter in order to analyze the varying tendency, ignoring the multi-value phenomenon and some useful information. The other is to give some values of the chaos parameters according to the comprehension of the authors, which is not convenient for the tendency analysis and therefore is not easy to manipulate for industrial application. The multi-value phenomenon is probably the main cause which results in the disagreement on the varying tendency of the chaos parameter under similar operating conditions [6, 28, 36]. In fact, the multi-value phenomenon of the chaos parameter for multiphase reactors has its intrinsic physical source [37]. Hence, the ordinary chaotic analysis that aims at estimating the concrete accurate value of chaos parameter and then analyzing the varying tendency is not enough to get well-accepted, easy to use results. And the estimating process of chaos parameter itself may be worth studying.

In the first part of this chapter, the bubbling mechanism in a gas–liquid bubble column with single orifice was studied with the signals of pressure fluctuations measured from a pressure transducer probe [3, 9]. The pressure transducer probe was placed in the bubble column close to the orifice mouth of gas to get enough strong signals of bubbling pressure. Thus, the pressure signals obtained in this way directly reflect the gas–liquid flow in the bubble column and the analysis results are more convenient for the industry applications. A local non-linear short-term prediction of pressure fluctuations was employed as a supplemental method to confirm the presence of the chaotic behavior and to unravel the non-linear characteristics of the system besides the time series plot, power spectrum, phase plot, Kolmogorov entropy and correlation dimension analyses. After non-linear prediction analysis, the predictability of the system was obtained.

In the second part of this chapter, the multi-value phenomenon of the correlation dimension was studied [3, 6, 10]. The time series of pressure fluctuations obtained from gas–liquid bubble columns was taken as the chaotic analysis signal. The mechanism of the multi-value phenomenon, its relationship with the flow behavior and possible application in the identification of flow regime and regime transition were investigated.

The nonlinear dynamics of other types of multiphase reactors were discussed in the third part of this chapter and at last some concluding remarks were summarized [3–5, 7, 8].

It is worth noting that the gas–solid fluidized bed is also an important type of reactor and several extensive reviews on the nonlinear analysis on this type of reactor have been done by several researchers [38–40]. They reviewed the applications of chaotic analysis to multiphase reactors. Examples were given by the characterization of regimes and regime transitions, scale up, control of the bubble pattern in the reactor to influence selectivity and conversion of chemical reactions, and the development of a tool to early detect agglomeration in fluidized beds. Hence, chaotic analysis of time series on gas–solid fluidized beds is not reviewed in this chapter. This work mainly focuses on the nonlinear investigations of liquid multiphase reactors. The chaotic dynamic characteristics of gas–liquid and gas–liquid–solid reactors are quite different from those of gas–solid reactors. On the other hand, the work emphasizes the calculation process and multi-value phenomena of chaos invariants.

References

1. R. Clift, J.R. Grace, M.E. Weber, *Bubbles, Drops and Particles* (Academic Press, New York, 1978)
2. L.S. Fan, K. Tsuchiya, *Bubble Wake Dynamics in Liquids and Liquid-Solid Suspensions* (Butterworths, Bosten, 1990)
3. M.Y. Liu (1998) Studies on the Chaos Hydrodynamic Characteristics of Multiphase Reactors. Dissertation, Tianjin University
4. M.Y. Liu, J.P. Wen, X.Y. Qin, Z.D. Hu, Local chaos characteristics in a self-aspirated reversed flow jet loop reactor. Trans. Tianjin Univ. **14**, 56–59 (1998)
5. M.Y. Liu, J.Y. Wu, Z.D. Hu, Chaos characteristics in a gas-liquid-solid three-phase magnetized fluidized bed. J. Chem. Eng. Chin. Univ. **13**, 476–480 (1999)
6. M.Y. Liu, Z.D. Hu, Chaos analysis of flow regime and regime transition in gas-liquid two-phase bubble columns. Eng. Chem. Metall. **21**, 37–43 (2000)
7. M.Y. Liu, Z.D. Hu, Chaos analyses of flow regime and regime transition in gas-liquid-solid three-phase fluidized beds. Chem. React. Eng. Technol. **16**(4), 363–368 (2000)
8. M.Y. Liu, J.H. Li, M. Kawauk, Application of the energy-minimization multi-scale method to gas-liquid-solid fluidized beds. Chem. Eng. Sci. **56**, 6805–6811 (2001)
9. M.Y. Liu, Z.D. Hu, Studies on the hydrodynamics of chaotic bubbling in a gas-liquid bubble column with a single orifice. Chem. Eng. Technol. **27**, 537–547 (2004)
10. M.Y. Liu, J.H. Li, Z.D. Hu, Multi-scale characteristics of chaos behavior in gas-liquid bubble columns. Chem. Eng. Commun. **191**, 1003–1016 (2004)
11. D.J. Tritton, C. Egdell, Chaotic bubbling. Phys. Fluids A **5**, 503–505 (1993)
12. L.J. Mittoni, M.P. Schwarz, R.D. La Nauze, Deterministic chaos in the gas inlet pressure of gas-liquid bubbling systems. Phys. Fluids A **7**, 891–893 (1995)
13. K. Nguyen, C.S. Daw, P. Chakka, M. Cheng, D.D. Bruns, C.E.A. Finney, M.B. Kennell, Spatio-temporal dynamics in a train of rising bubbles. Chem. Eng. J. **64**, 191–197 (1996)
14. H.Z. Li, Y. Mouline, L. Choplin, N. Midoux, Chaotic bubble coalescence in non-Newtonian fluids. Int. J. Multiphas Flow **23**, 713–723 (1997)
15. M.C. Ruzicka, J. Drahos, J. Zahradnik, N.H. Thomas, Intermittent transition from bubbling to jetting regime in gas-liquid two phase flows. Int. J. Multiphas Flow **23**, 671–682 (1997)
16. W. Luewisutthichat, A. Tsutsumi, K. Yoshida, Chaotic hydrodynamics of continuous single-bubble flow systems. Chem. Eng. Sci. **52**, 3685–3691 (1997)
17. A. Tufaile, J.C. Sartorelli, Chaotic behavior in bubble formation dynamics. Phys. A **275**, 336–346 (2000)
18. A. Tufaile, J.C. Sartorelli, Henon-like attractor in air bubble formation. Phys. Lett. A **275**, 211–217 (2000)
19. A. Tufaile, M.B. Reyes, J.C. Sartorelli, Explosion of chaotic bubbling. Phys. A **308**, 15–24 (2002)
20. R. Mosdorf, M. Shoji, Chaos in bubbling-nonlinear analysis and modeling. Chem. Eng. Sci. **58**, 3837–3846 (2003)
21. S.U. Sarnobat, S. Rajput, D.D. Bruns, D.W. DePaoli, C.S. Daw, K. Nguyen, The impact of external electrostatic fields on gas–liquid bubbling dynamics. Chem. Eng. Sci. **59**, 247–258 (2004)
22. H.Z. Li, X. Frank, D. Funfschilling, P. Diard, Bubbles' rising dynamics in polymeric solutions. Phys. Lett. A **325**, 43–50 (2004)
23. X. Frank, H.Z. Li, Route to chaos in the rising dynamics of a bubble chain in a polymeric fluid. Phys. Lett. A **372**, 6155–6160 (2008)
24. J.T. Cieslinski, R. Mosdorf, Gas bubble dynamics - experiment and fractal analysis. Int. J. Heat Mass Tran. **48**, 1808–1818 (2005)
25. P. Garstecki, M.J Fuerstman, G.M Whitesides Nonlinear dynamics of a flow-focusing bubble generator: an inverted dripping faucet. Phys. Rev. Lett. **94**, 234502-1–234502-4 (2005)

26. R. Mosdorf, T. Wyszkowski, Experimental investigations of deterministic chaos appearance in bubbling flow. Int. J. Heat Mass Tran. **54**, 5060–5069 (2010)
27. M. Ruzicka, J. Drahos, J. Zahradnik, N.H. Thomas, Structure of gas pressure signal at two-orifice bubbling from a common plenum. Chem. Eng. Sci. **55**, 421–429 (2000)
28. H.M. Letzel, J.C. Schouten, R. Krishna, C.M. van den Bleek, Characterization of regimes and regime transitions in bubble columns by chaos analysis of pressure signals. Chem. Eng. Sci. **52**, 4447–4459 (1997)
29. R. Femat, J.A. Ramirez, A. Soria, Chaotic flow structure in a vertical bubble column. Phys. Lett. A **248**, 67–79 (1998)
30. W. Chen, T. Hasegawa, A. Tsutsumi, K. Otawara, Y. Shigaki, Generalized dynamic modeling of local heat transfer in bubble columns. Chem. Eng. J. **96**, 37–44 (2003)
31. F. Franca, M. Acikgoz, R.T. Lahey, A. Clausse, The use of fractal techniques for flow regime identification. Int. J. Multiphas Flow **17**, 545–552 (1991)
32. A.I. Karamavruc, N.N. Clark, Local differential pressure analysis in a slugging bed using deterministic chaos theory. Chem. Eng. Sci. **52**, 357–370 (1997)
33. D. Bai, E. Shibuya, N. Nakagawa, K. Kato, Fractal characteristics of gas-solids flow in a circulating fluidized bed. Powder Technol. **90**, 205–212 (1997)
34. D. Bai, A.S. Issangya, J.R. Grace, Characteristics of gas-fluidized beds in different flow regimes. Ind. Eng. Chem. Res. **38**, 803–811 (1999)
35. R. Kikuchi, T. Yano, A. Tsutsumi, K. Yoshida, M. Punchochar, J. Drahos, Diagnosis of chaotic dynamics of bubble motion in a bubble column. Chem. Eng. Sci. **52**, 3741–3745 (1997)
36. Y. Kang, Y.J Cho, K.J Woo, K.I Kim, S.D Kim Bubble properties and pressure fluctuations in pressurized bubble columns. Chem. Eng. Sci. **55**, 411–419 (2000)
37. B.R. Bakshi, H. Zhong, P. Jiang, L.S. Fan, Analysis of flow in gas-liquid bubble columns using multi-resolution methods. Trans. IchemE Part A **73**, 608–614 (1995)
38. F. Johnsson, R.C. Zijerveld, J.C. Schouten, C.M. van den Bleek, B. Leckner, Characterization of fluidization regimes by time-series analysis of pressure fluctuations. Int. J. Multiphas Flow **26**, 663–715 (2000)
39. J.R. van Ommen, S. Sasic, J. van der Schaaf, S. Gheorghiu, F. Johnsson, M.O. Coppens, Time-series analysis of pressure fluctuations in gas–solid fluidized beds – A review. Int. J. Multiphas Flow **37**, 403–428 (2011)
40. C.M. van den Bleek, M.O. Coppens, J.C. Schouten, Application of chaos analysis to multiphase reactors. Chem. Eng. Sci. **57**, 4763–4778 (2002)

Chapter 2
Experimental

2.1 Apparatus and Set-ups

Figure 2.1 shows the experimental apparatus and set-up for gas–liquid bubbling system with single orifice [1–2].

The experiments were conducted in a Plexiglas column with an inner diameter of 0.07 m and a height of 1.55 m. Nitrogen was employed as the gas phase. Distilled water was used as the liquid phase and its temperature is typical about 302 K. The gas bubbles were injected consecutively into the bubble column with a static liquid height of 0.80 m through a horizontal orifice with a diameter of 1.2 mm, at the bottom and in the center of the column. The pressure signal near the orifice was measured relative to atmosphere with a high accuracy pressure transducer (IC-Sensor, 1220-2psid). The response time of this pressure transducer was about 10^{-5} s. The transducer probe was placed typically about 0.01 m above the orifice and 0.01 m away from the axis of symmetry of the bubble column. The bubbling process was the type of constant flow rate. The gas flow rate was set by a mass flow controller.

The experimental equipment and setup for the gas–liquid multi-orifice bubbling system is similar to that of the single orifice [1, 3]. Tap water and air were used as the liquid and gas phases, respectively. The superficial gas velocity was varied over the range of 0.177–9.929 cm/s. The superficial liquid velocity was zero, and the static liquid height was 0.88 m. The time series of pressure fluctuations were measured relative to atmosphere. The sensor probe was placed in the axial center of the column at an axial position of 0.49 m from the distributor of the bubble columns.

2.2 Data Acquisition

For the system of single-orifice gas–liquid bubbling, digitized time series of pressure fluctuations of more than 10 s were recorded with a sample frequency of 3,000 Hz when the operation of the bubble column was in a steady state. A typical

M. Liu and Z. Hu, *Nonlinear Analysis and Prediction of Time Series in Multiphase Reactors*, SpringerBriefs on Multiphase Flow, DOI: 10.1007/978-3-319-04193-3_2, © The Author(s) 2014

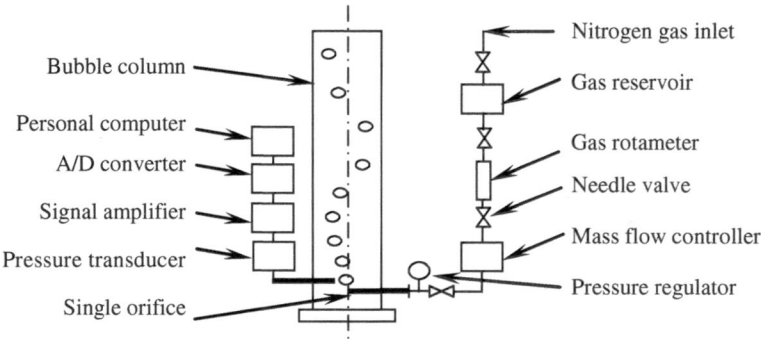

Fig. 2.1 Experimental apparatus and set-up of single-orifice bubbling system

sample size was 32,768 points. The data were logged into a personal computer via a signal amplifier and an A/D converter with 12-bit resolution and accuracy better than 99.97 % [1].

The pressure signals measured from the multiphase systems are subject to noise, especially for the gas–liquid bubble columns. The non-linear analysis is very sensitive to the noise of the signals. In gas–solid fluidized beds, the signals are transmitted to the probes by the suspension of particles in gas phase, and can be dampened considerably. However, in gas–liquid bubble columns, near all the pressure fluctuations can be transmitted very well to the pressure probes by the liquid phase. The noise can significantly increase the correlation dimension. Hence, some measures must be taken to reduce the noise even though the issue of noise reduction is still a worldwide challenge. In this work, the signals were filtered off-line with a frequency of 60 Hz by using a low-pass algorithm to reduce low amplitude, high frequency noise, and at the same time to ensure that the spectra of interest of the hydrodynamic signal were still captured. This treatment of noise reduction is similar to that of most studies on the chaos analysis of signals in a gas–liquid bubble column [4].

For the system of multi-orifice gas–liquid bubbling, a total of 32,768 data points were acquired for each pressure signal at a sample frequency of 300 Hz. Other data treatments are the same as those of the system of single-orifice gas–liquid bubbling.

References

1. M.Y. Liu, Studies on the chaos hydrodynamic characteristics of multiphase reactors, Dissertation, Tianjin University, 1998
2. M.Y. Liu, Z.D. Hu, Studies on the hydrodynamics of chaotic bubbling in a gas-liquid bubble column with a single orifice. Chem. Eng. Technol. **27**, 537–547 (2004)

3. M.Y. Liu, J.H. Li, Z.D. Hu, Multi-scale characteristics of chaos behavior in gas-liquid bubble columns. Chem. Eng. Commun. **191**, 1003–1016 (2004)
4. H.M. Letzel, J.C. Schouten, R. Krishna, C.M. van den Bleek, Characterization of regimes and regime transitions in bubble columns by chaos analysis of pressure signals. Chem. Eng. Sci. **52**, 4447–4459 (1997)

Chapter 3
Analysis Tools of Time Series Data

Details of the theory of chaotic analysis and prediction of time series signals and methods are available in literature and only a brief description is given here [1–10].

3.1 Time Series

The analysis of the signals in time domain is the first step and can obtain some useful preliminary information about the system hydrodynamics. The simplest analysis in time domain is to plot a sequence of data points of the measured signal. This gives a qualitative description of the time scale or of the complexity of the flow. For example, the periodic bubbling process can be easily identified by observing the signals in time domain.

3.2 Power Spectrum

The power spectrum shows how the energy of a signal is distributed over the frequencies. It reveals characteristic time scales of the hydrodynamics and provides preliminary identification of dynamic changes of the system. For periodic and multi-periodic signals, the power spectra exhibit discrete frequencies. If the power spectrum exhibits broadband noise at low frequencies, it is an indication of a complex signal. But, other effective analysis tools are needed in order to further identify and explain the complex behavior.

3.3 Attractor Reconstruction and Phase Plot

Nearly all of the methods of non-linear analysis of signals are on the basis of the reconstruction of an attractor of the dynamic evolution of the system in phase space or state space. It is possible to represent any physical system by a plot in its

M. Liu and Z. Hu, *Nonlinear Analysis and Prediction of Time Series in Multiphase Reactors*, SpringerBriefs on Multiphase Flow, DOI: 10.1007/978-3-319-04193-3_3, © The Author(s) 2014

phase space, which is an imaginary space with a few axes, each representing a
state variable of the system. Every single state of the system corresponds to a point
in the phase space. In a phase space, the curve of each state point at consecutive
time steps is called a trajectory. The part of the phase space to which the trajectories converge is called the attractor of the system. The attractor can be reconstructed by means of only one characteristic variable such as the time series of
pressure fluctuations, using the method of delay coordinates. The constructed
phase space can be used to measure the geometric and dynamic characteristics of
an attractor.

In the study of this chapter, Takens' [11] embedding theorem was applied to
reconstruct attractors. A set of time series data of pressure fluctuations was used to
generate a m-dimensional state vector.

If $\{x_1, x_2, x_3, \ldots \ldots, x_M\}$ is a measured scalar time series signal and X_i a m-
dimensional reconstructive vector, which is given by

$$X_i = \left(x_i, x_{i+\tau}, \ldots, x_{i+(m-1)\tau/\Delta t} \right)^T (i = 1, 2, \ldots \ldots, N) \tag{3.1}$$

where τ is the time delay, m is the embedding dimension and N is the length of
restructured vector X_i

The structure of the attractor was studied in a phase space. This plot is called a
phase plot of attractors, which can give some qualitative information about the
dynamics of the system. For example, a periodic signal shows up as a limit cycle
on the phase plot, whereas a chaotic signal exhibits a very complex and rich
structure of the attractor.

3.4 Correlation Dimension

The correlation integral, $C_{(r)}$, is defined as

$$C_{(r)} = \lim_{N \to \infty} \frac{2}{N(N-1)} \sum_{\substack{i,j=1 \\ i \neq j}}^{N} H(r - |X_i - X_j|) \tag{3.2}$$

where $H(x)$ is the Heaviside step function and $|X_i - X_j|$ is the scalar distance
between points on the solution trajectory and r is the radius of the hyper-sphere. It
is found that the correlation integral $C_{(r)}$, has a power law dependence on r, that is

$$C_{(r)} \sim r^{D_2} \tag{3.3}$$

where D_2 is called the correlation dimension and is a measure for the fractal
dimension of the attractor. Since the correlation dimension displays spatial correlation between the points on the attractor, it is also a measure of the spatial
homogeneity in the phase space [12]. In principle, the correlation dimension is
obtained by fitting the slope of a log-log plot of correlation integral $C_{(r)}$, versus

radius of the hyper-sphere, r. With the increase of m, the slope of the plot will become independent of m and converge to a saturation value, which is chosen as the correlation dimension of the signal analyzed.

3.5 Kolmogorov Entropy

The Kolmogorov entropy K, is a quantitative measure of the rate of information loss of the system dynamics. The information loss in a chaotic system arises from the exponential divergence of very close trajectories. For a periodic system, its value is zero. For a pure random system, its value is positive infinity, making it impossible to predict the state of the system even after a differential time step. For the case of a chaotic system, it is finite and positive. For a practical or real dynamic system, time series data obtained from the experiments will be contaminated inevitably by noise. In this situation, an absolute zero value of K can't be obtained by the chaotic calculation since a periodic motion in a strict sense can't be found in a real system. Similarly, the positive infinity of K can't be obtained by the non-linear calculation because a pure random motion can't be found in a real system. In this chapter, a K value of less than 5 bits/s (zero rigorously) is considered as an indicator of a periodic motion, and a K value of higher 80 bits/s (positive infinity theoretically) is considered as an indicator of a random motion.

Grassberger and Procaccia [7] proposed an easier method to estimate the K from the correlation integral. Using the relationship between the joint probability and the correlation integral, they approximated order-2 Renyi entropy [8] $K_{2,m}$, by

$$K_{2,m} = \frac{1}{\tau} \ln \frac{C_{m(r)}}{C_{m+1(r)}} \tag{3.4}$$

The $K_{2,m}$ is numerically close to the actual value of K.

3.6 The Largest Lyapunov Exponent

Lyapunov exponents quantify the exponential divergence of initially close state-space trajectories and estimate the degree of chaos in systems. The presence of a positive characteristic exponent indicates chaos. Furthermore, in many applications it is sufficient to calculate only the largest Lyapunov exponent (LLE or λ_1) to detect the presence of chaos in a dynamical system. For example, consider two trajectories with nearby initial conditions on an attracting manifold. When the attractor is chaotic, the trajectories diverge, on average, at an exponential rate characterized by the largest Lyapunov exponent. The largest Lyapunov exponent was calculated from an experimental time series based on the programs developed by Wolf et al. [9].

$$\lambda_1 = \frac{1}{t_n - t_o} \sum_{k=1}^{n} \log_2 \frac{L'(t_k)}{L(t_{k-1})} \tag{3.5}$$

3.7 Local Non-Linear Short-Term Prediction

It is shown that the local non-linear short-term prediction of a signal is an effective diagnostic tool to identify chaos [10]. The ability to predict successfully with this method can be the strongest test of whether or not low-dimensionality chaos is present, and also an effective way of distinguishing chaos from random. An algorithm exploiting non-linear short-term prediction of a signal is developed following the approach introduced by a few authors [10]. It is based on the idea that for a chaotic time series the predictability is limited and the accuracy of the prediction falls off when increasing the prediction-time interval, whereas for uncorrelated noise or random signal, the predictability is zero and the accuracy is independent of prediction time interval. Obviously, for a noise-free periodic signal, the predictability is infinite.

The first step of the non-linear short-term prediction is to reconstruct the data of a time series into an attractor in a phase space according to the embedding theorem. The next step is to assume a functional relationship between the current state $X_{(t)}$, and the future state, $X_{(t+T)}$.

$$X_{(t)} = gX_{(t+T)} \tag{3.6}$$

We want to find a predictor G that approximates g. There are several possible approaches. One of the effective approaches is the local approximation, using only nearby states to make predictions. The simplest method to build a local predictor is approximation by nearest neighbor, i.e., zeroth-order approximation. In this case, the number of the nearby states k equals to one. A superior one is the first-order approximation, with taking k greater than embedding dimension m. In order to facilitate the comparison of results, we simply build the database from the first part of the time series, and hold it fixed as we make predictions on the remainder.

To evaluate the accuracy of the local non-linear short-term prediction, the correlation coefficients between predicted and experimentally measured values are calculated. The decrease in the correlation coefficient when increasing the number of predicted points is a feature of chaos if the noise in the signal is basically reduced.

It is noted that the estimations of chaotic invariants are very sensitive to the parameter choices and datum noise. Hence, several aspects should be taken into consideration when estimating the correlation dimension and Kolmogorov entropy [1] using the signals measured from practical physical systems. As mentioned above, reducing noise is a measure to ensure the calculation accuracy. On the other hand, datum length, time delay and embedding dimension etc. should be chosen carefully. Most studies have shown that 3,000–10,000 points are sufficient for the multiphase flow systems. In this article, reasonable estimates of the correlation dimension can be reached when $M = 5,000$ and $N = 4,500$. The time delay τ should be chosen carefully in order to reach the linear independence of the phase space coordinates. Although any time delay will be acceptable according to the embedding theorem, some choice rule should be observed from the practical point of view. Generally, one can take the first zero-crossing time, the time of the first minimum of the autocorrelation function (in our case, $\tau \approx 0.11$ s), or the time of the first minimum of the mutual information function as a good hint of time delay. The embedding dimension m, should be large enough to encompass the complete reconstructed attractor. There are basically four methods to search a necessary embedding dimension: singular-value decomposition of the sample covariance matrix; "saturation" with dimension of some system invariant (in our case); the method of false nearest neighbors; the method of true vector fields. In the study of this chapter, a constant slope is reached at $m \approx 14$ in most case, and a higher value $m \approx 16$ has been used.

Even though the datum parameters such as the datum length is somewhat limited to estimate accurately the Kolmogorov entropy, correlation dimension and the largest Lyapunov exponent, and a positive value of Kolmogorov entropy or the largest Lyapunov exponent is not sufficient to conclude the existence of chaotic behavior in experimental systems according to the view of the general non-linear analysis theory. However, they can be seen as the good indicators. What's more, the variation tendencies of chaotic invariants with the operation conditions such as the gas flow rate or velocity are of more important than their accurate values.

References

1. M.Y. Liu, Studies on the chaos hydrodynamic characteristics of multiphase reactors. Dissertation, Tianjin University, (1998)
2. M.Y. Liu, Z.D. Hu, Studies on the hydrodynamics of chaotic bubbling in a gas-liquid bubble column with a single orifice. Chem. Eng. Technol. **27**, 537–547 (2004)
3. M.Y. Liu, J.H. Li, Z.D. Hu, Multi-scale characteristics of chaos behavior in gas-liquid bubble columns. Chem. Eng. Commun. **191**, 1003–1016 (2004)
4. F. Johnsson, R.C. Zijerveld, J.C. Schouten, C.M. van den Bleek, B. Leckner, Characterization of fluidization regimes by time-series analysis of pressure fluctuations. Int J Multiphas Flow **26**, 663–715 (2000)
5. T. Mullin, *The nature of chaos* (Clarendon Press, Oxford, 1993)
6. R.C. Hilborn, *Chaos and non-linear dynamics: an introduction for scientists and engineers* (Oxford University Press, New York, 1994)
7. P. Grassberger, I. Procaccia, Estimation of the Kolmogorov entropy from a chaotic signal. Phys. Rev. A **28**, 2591–2593 (1983)
8. A. Renyi, *Probability theory* (North-Holland, Armsterdam, 1970)
9. A. Wolf, J.B. Swift, H.L. Swinney, J.A. Vastano, Determining Lyapunov exponents from a time series. Physica D **16**, 285–317 (1985)
10. J.D. Farmer, J.J. Sidorwich, Predicting chaotic time series. Phys. Rev. Lett. **59**, 845–848 (1987)

11. F. Takens, Detecting strange attractors in turbulence. *Lecture notes in mathematics*, vol 898 (Springer, Berlin, 1981), p. 366
12. Y. Kang, K.J. Woo, M.H. Ko, S.D. Kim, Particle dispersion and pressure fluctuations in three-phase fluidized beds. Chem. Eng. Sci. **52**, 3723–3732 (1997)

Chapter 4
Results and Discussion

4.1 Single-Orifice Bubbling Mechanism

4.1.1 Time-Domain and Power Spectrum

The basic patterns of pressure fluctuations and their power spectra when increasing the gas flow rate are shown in Fig. 4.1 [1, 2].

When gas flow rate Q, is low, single-bubble formation and sometimes double-bubble release were observed. The bubble diameter is about 0.004 m estimated by the naked eyes [1]. For example, when Q is 73 ml/min (e.g. 1.217×10^{-6} m³/s), the period-1 bubbling with a bubbling frequency of 25.8 was identified, indicating that the pressure signal is periodic. The bubbling frequency 25.8 was obtained from the time series and power spectrum figures, as shown in Fig. 4.1a and i. In this bubbling process, each spherical bubble moved upwardly and freely. The bubbling flow is called periodic bubbling.

When increasing Q, the bubbling frequency, the size and velocity of bubbles increase. The bubble diameters vary from 0.005 to 0.006 m, and a bubbling frequency of 42.8 was reached. In the range of these gas flow rates, the bubbling system begins to lose its stability. Figure 4.1b and j shows the time series and power spectrum of pressure fluctuation at Q of 192 ml/min, respectively. It can be seen from these figures that the signal is slightly modulated by a low frequency. This drift phenomenon in the signals is due to the bubble-bubble interactions and the circulated fluid in the column. The rise velocity of one bubble increases due to the existence of the wake of its leading bubble. Even though the process is unstable, the sizes of the bubbles are equal, which can be concluded from the amplitude of the pressure signal. The further increase of Q enhances the system unstableness. And the bubbles rose in a continuous chain of bubbles. The difficulty occurs when trying to distinguish single and multiple bubbling. The size of the bubble is even, and the diameter is about 0.006–0.008 m. The non-uniformity of the system appeared both in amplitude and frequency of the pressure signal. This

M. Liu and Z. Hu, *Nonlinear Analysis and Prediction of Time Series in Multiphase Reactors*, SpringerBriefs on Multiphase Flow, DOI: 10.1007/978-3-319-04193-3_4, © The Author(s) 2014

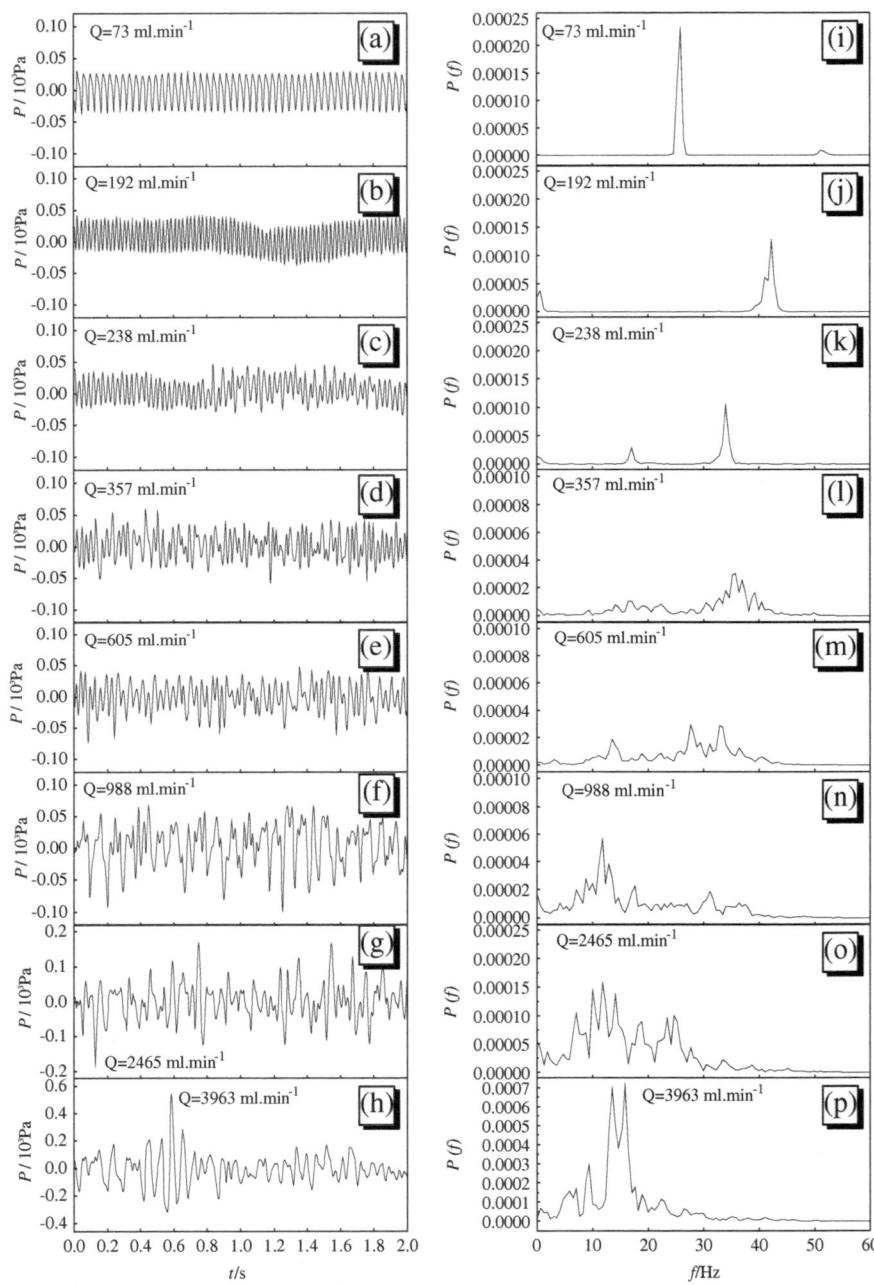

Fig. 4.1 Time series and power spectra of pressure signals vs. gas flow rates [1, 2]. **a–h**, time series; **i–p**, power spectra

case is illustrated in Fig. 4.1c and k, and the gas flow rate Q is 237 ml/min with a bubbling frequency of about 34.2. The decrease of bubbling frequency with an increase of gas flow rate may result from the larger bubble diameter due to the coalescence of smaller bubbles at the tip of the orifice. Besides the modulation of a low frequency, the power spectrum begins to exhibit some broadband character-istics, as shown in Fig. 4.1j and k together with different sharp peaks as shown in Fig. 4.1k, which may denote the onset of chaos behavior. Further evidence can be found from the Kolmogorov entropy analysis. This unstable bubbling flow is called primary chaos bubbling.

When Q is high, the break-up and coalescence of the bubbles occured in the column. The bubble size is uneven, basically ranging from 0.002 to 0.025 m. The number of smaller bubbles is large, but the number of larger bubbles with a diameter of more than 0.025 m is small. Some interactions in the gas–liquid two-phase flow were excited and the bubbling process is irregular. The power spectra of Q equaling 357, 605 and 988 ml/min display broadened spectral lines with no discernible dominant frequencies as shown in Fig. 4.1l–n. This indicates qualita-tively a character of chaotic bubbling. The time series are stable, as shown in Fig. 4.1d–f. When Q is 357 ml/min, the shapes of most bubbles are elliptical, with an equivalent diameter of about 0.008 m. Meanwhile, a number of small bubbles with a diameter of 0.003 m were found. These results indicate that the break-up and coalescence of the bubbles occured. When Q is 605 ml/min, the break-up and coalescence were more obvious. When Q is 988 ml/min, the size distribution of the bubbles is wide, and a swarm of bubbles or the large bubbles with a diameter of 0.025 m were often observed. The bubbling process approaches jetting. The flow of this bubbling is called advanced chaotic bubbling.

When Q is very high, the break-up and coalescence of the bubbles happened frequently. Nearly all the interactions in the gas–liquid two-phase flow were excited and the bubbling process is very irregular. The size of the bubbles increases and the bubble diameter can approach the diameter of the bubble col-umn. This kind of large bubble is called a slug bubble, and the flow is called slug flow. The slug bubble can be seen from the time series of the pressure signal as shown in Fig. 4.1h. The power spectra of Q equaling 2465 and 3963 ml/min also display broadened spectral lines with no discernible dominant frequencies as shown in Fig. 4.1o and p, especially in figures of half-log coordinate system (not shown here), which indicates that the process is in a high-level chaos [3]. The process characteristic is more like jetting when considering its predictability. So, this flow is called jetting or random bubbling.

These analysis results are helpful for the identification of the system's non-linear hydrodynamics. However, it is difficult to get affirmative evidence of chaos bubbling and to extract more qualitative and quantitative hydrodynamic infor-mation only from the analyses of time-domain and power spectrum.

4.1.2 Phase Plane Plot

A phase plot can describe the attractor structure of a bubbling hydrodynamic
system. Figure 4.2 shows the phase plots of pressure fluctuations under different
gas flow rates. The processes of regular bubbling can be identified from these
phase plots. When Q equals to 73 ml/min, the phase plot shows a regular ring
symbolizing a mono-frequency phenomenon, as shown in Fig. 4.2a. Even though
the limited ring is not rigorous from a point of view of the theoretical definition
due to the irregularity of the experimental signal, the periodic bubbling process is
basically tested. When Q is 192 ml/min, the phase plot is a two-dimension ring
area as shown in Fig. 4.2b, which indicates a feature of a periodicity bubbling
process with a low frequency modulation. When Q is high, every phase plot
reveals a very fine structure of the attractor with the appearance of many rings that
characterize complex bubbling hydrodynamics as shown in Fig. 4.2c–h. The formation of each ring in the phase plot introduces a new independent frequency.
Thus, the ring structures of these phase plots were not clearly seen due to the
feature of broadened spectrum appearing in the power spectrum. But the structure
is rich and not in disorder. Under these conditions, the chaotic bubbling is identified qualitatively.

4.1.3 Kolmogorov Entropy and Correlation Dimension

The above analyses are qualitative and more advanced quantitative analyses, such
as Kolmogorov entropy and correlation dimension analyses, are needed in order to
give more convincing results. Figure 4.3a and b shows the variations of K and D_2
with Q, respectively.

It can be seen from Fig. 4.3 that gas–liquid flow regime can clearly and
quantitatively be identified by the analyses of the non-linear invariants. The variation tendencies of the two invariants are nearly the same and the change of
K with Q is more obvious. However, the variation of the standard deviation of the
signal, δ, or the size of the attractor of the bubbling system with Q is relatively flat,
which can be seen from Fig. 4.4.

It is noted that even though the curve of standard deviation of pressure signal
with gas flow rate is flat (Fig. 4.4), the curves of Kolmogorov entropy and correlation dimension with gas flow rate are not flat and there are several inflection
points in the curves (Fig. 4.3). The main reason is that the nonlinear analysis
method utilizes fully the multi-dimensional information hidden in the time series
data by phase space reconstruction technique, while the standard deviation analysis (a linear analysis method) only uses the one-dimensional information of
measured signal. That is why we should not only use the traditional linear analysis
method (for example, standard deviations, etc.) to study the bubbling dynamic
behavior.

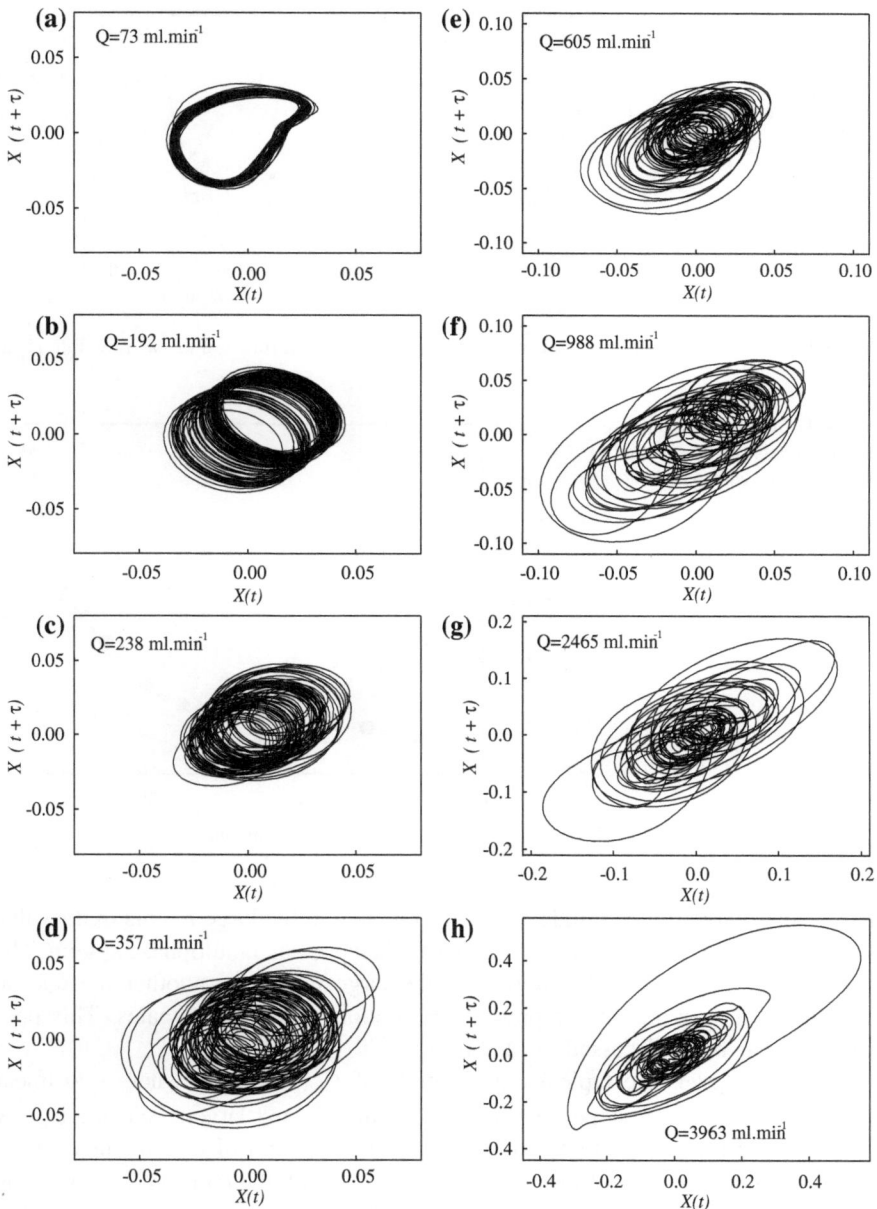

Fig. 4.2 Phase plots of pressure fluctuations under different gas flow rates [1, 2]

In the estimation of Kolmogorov entropy and correlation dimension, two values were obtained at a given Q for the low or high flow rates. One is small in magnitude and was estimated in the middle r in the log-log plot of $C_{(r)}$ versus r.

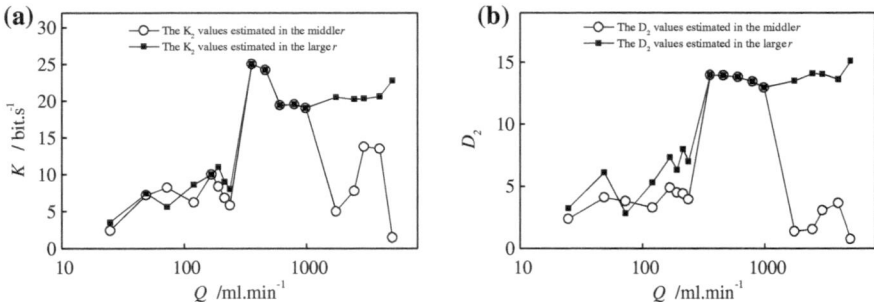

Fig. 4.3 **a** Kolmogorov entropy or **b** correlation dimension as a function of gas flow rate [1, 2]

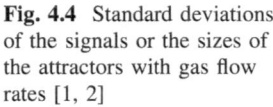

Fig. 4.4 Standard deviations of the signals or the sizes of the attractors with gas flow rates [1, 2]

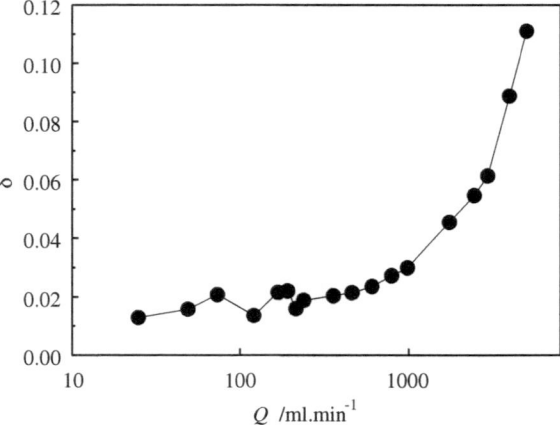

The other is large in magnitude and was estimated in the large r. This multi-value phenomenon may result from the multi-scale behavior in a multiphase system. The smaller one may reflect the motion of the large bubble or another meso-scale motion, the larger one may represent the motion of the other scales. This phenomenon has been discussed in literature [4].

Figure 4.5 gives the typical log-log plots of $C_{(r)}$ versus r, where two linear sections with different slopes can be seen for middle and large r. When r is very small, the third linear section can be identified in the plot of $C_{(r)}$ versus r. However, the slope variation of the third linear section with the embedding dimension is not regular, which can be seen from Fig. 4.5d. Hence, it was not considered in this work. It can also be found from Fig. 4.5 that the curve shapes in the log-log plots are different for distinct bubbling processes, which may be a potential and effective identification method of bubbling flow regimes.

When at a low Q, two Kolmogorov entropy values estimated in different r are low but positive in magnitude and their variation tendencies with Q are nearly the same, as shown in Fig. 4.3a. These results show that the bubbling dynamics at low

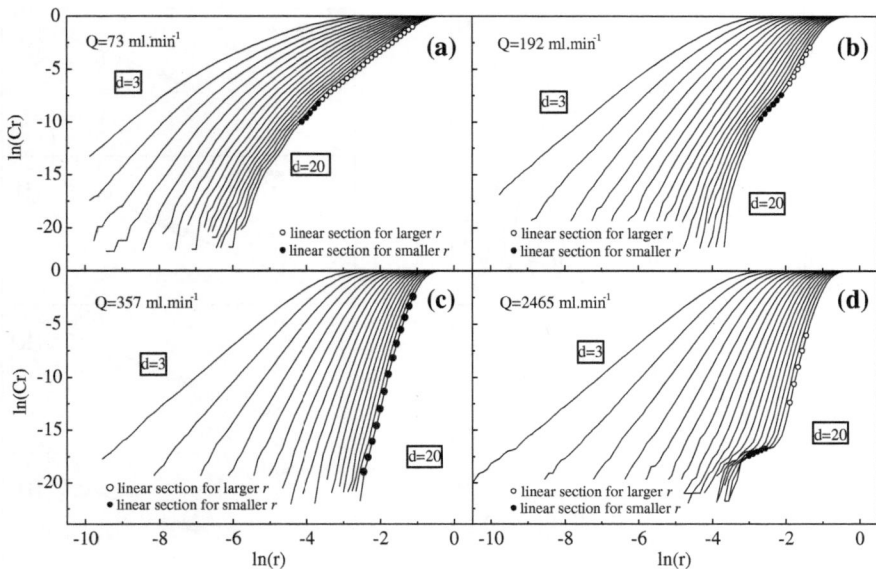

Fig. 4.5 Typical log-log plots of correlation integral versus r at different gas flow rates [1, 2]

gas flow rates in actual system are not rigorously periodic, even for the periodic bubbling system. For example, when $Q = 73$ ml/min, a periodic bubbling process is identified by time series, power spectrum and phase plot analyses and its K should be zero. However, the K is about 5.6 or 8.2 bits/s, not equaling to zero.

When in unstable bubbling operations, with Q ranging from 168 to 238 ml/min, the value of K first increases and then decreases slightly. The first increase of K indicates the onset of the chaotic bubbling and the decrease of K may result from a self-organization of the bubbling system due to the bubble-bubble interactions. This sudden drop of chaotic quantities in the vicinity of the transition point was also reported by Ruzicka et al. [5].

When increasing Q further, only one value of K was obtained for a given Q, and an abrupt increase of value of K is observed, indicating that the bubbling process begins to undergo another re-organization. The gas flow rate of this re-organization process varies from 357 to 988 ml/min. The value of K first increases and then drops down, showing that the bubbling dynamics change from a primary chaos to a high degree of chaos. And in this range of gas flow rates, the break-up and coalescence of bubbles dominate the bubbling process. This is the second drop in K value.

When Q exceeds 988 ml/min, the value of K calculated in larger r increases again slightly, and the jetting or random bubbling is established. Meanwhile, the value of K calculated in the middle r drops down sharply even though with some fluctuations. This may reflect the motion of slug bubble in the system because in these operation conditions, a few slug bubbles often go through the column with a low frequency motion. This is the third drop of K value corresponding to the middle r.

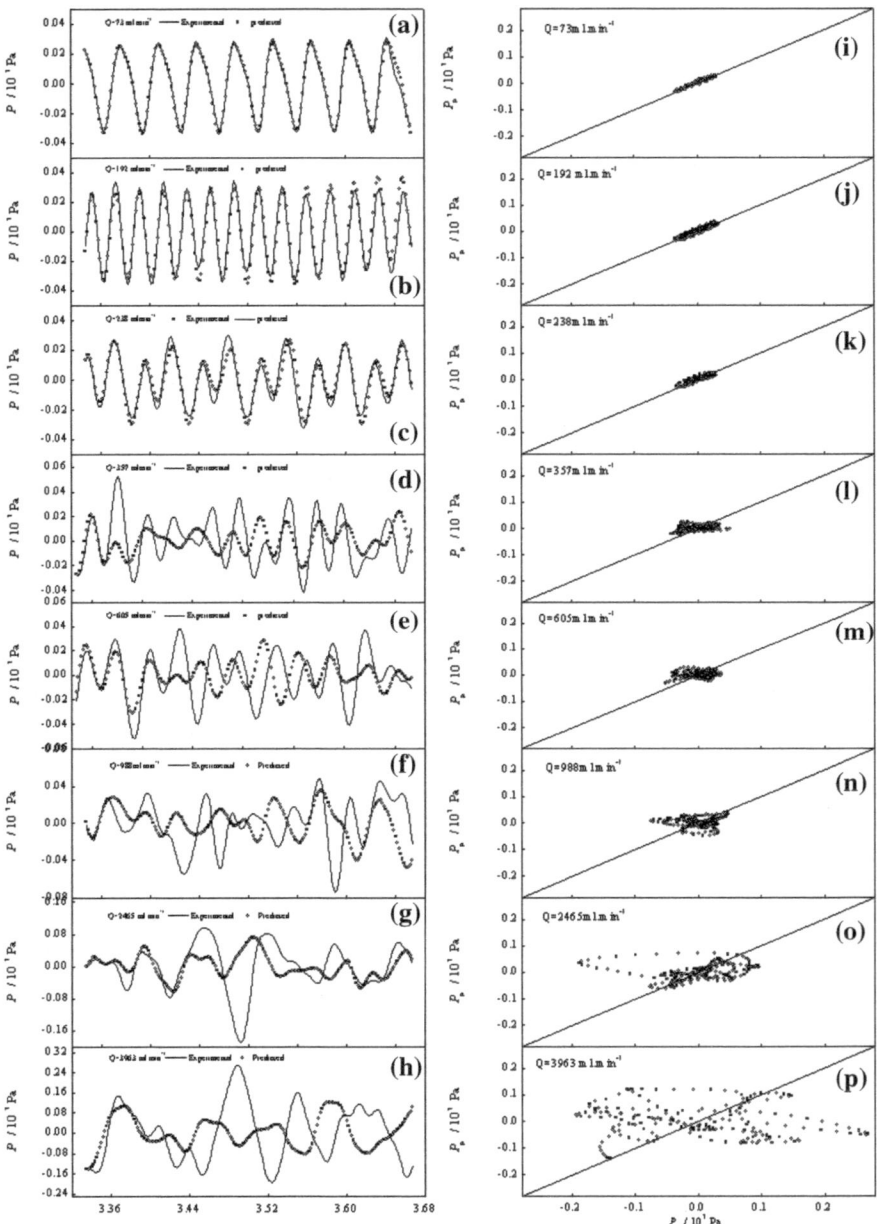

Fig. 4.6 Comparisons between predicted and experimental values of pressure fluctuations at different gas flow rates of bubbling process [1, 2]. **a–h**, comparisons of time series; **i–p**, comparisons of predicted and experimental values

It is obvious that K is a sensitive indicator of bubbling hydrodynamics. Similar results can be drawn from the analysis of variation tendency of D_2, which will not be discussed in detail. Considering the parameter sensitivity and the uncertainty of the estimations of K and D_2, more convincing and objective chaos diagnostic methods are needed to support the above points.

4.1.4 Local Non-Linear Short-Term Prediction

The local non-linear short-term prediction tool was used as a supplemental method for the chaos detection in the bubbling process. Figure 4.6 shows the typical comparisons between predicted and experimental values of pressure signals. Figure 4.6a–h shows the relationships between predicted and experimental time series. Figure 4.6i–p shows the comparisons of predicted and experimental values. When predicting, the time series data are re-sampled with a frequency of 600 Hz. The beginning 2000 points in the signal were used as a basis for making predictions for each of the later points. The prediction time is 0.33 s for each signal. The embedding dimension, m, and time delay, τ, are different for distinct gas flow rates. Generally, the prediction results are not very sensitive to the value of these parameters. In this work, the embedding dimensions were varied within the range $3 \leq m \leq 9$, and the time delays were varied from Δt to $30\Delta t$.

When the gas flow rate is low, for example, Q is 73 ml/min, a long prediction time with a high accuracy can be achieved, as shown in Fig. 4.6a–i. In Fig. 4.6i, the points are all on the diagonal, showing that the predicted and experimental values are very close. The good predictability indicates that the bubbling process at this gas flow rate is basically periodic. The length of the accuracy prediction time is about 0.3 s.

When increasing Q, the length of prediction time reduces. This can be seen from Fig. 4.6b–j, or from Fig. 4.6c–k. At these gas flow rates, the length of prediction time is shortened a little and the points in Fig. 4.6j or k are to a degree scattered around the diagonal. This is the characteristic of a primary chaos bubbling process.

When Q increases further, the length of the prediction time is further shortened. This can be seen from Fig. 4.6d and l, e and m, and f and n. For example, the time is about 0.04 s when Q equals 357 ml/min. After the prediction time, the ability to predict with high accuracy is lost all of a sudden. The points in Fig. 4.6l–n are scattered obviously. In these operation conditions, the bubbling is in an advanced chaos.

When Q is very high, the ability to predict with high accuracy is lost greatly. For instance, when Q is 2465 ml/min, the length of accuracy prediction time is shortened greatly, which is about 0.01 s, as shown in Fig. 4.6g and o. When Q is 3963 ml/min, the prediction time is less than 0.01 s, as shown in Fig. 4.6h and p. The points in Fig. 4.6p are scattered seriously. This is the characteristic of jetting or a random bubbling process.

Fig. 4.7 Correlation
coefficients between
predicted and experimental
values versus the number of
the predicted points [1, 2]

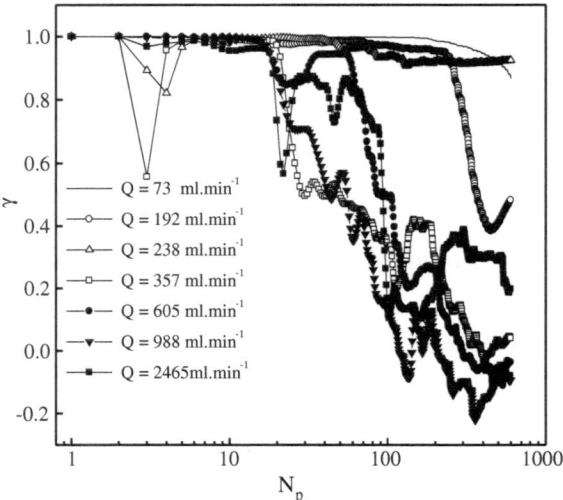

Figure 4.7 shows the correlation coefficients between predicted and experimental values versus the number of predicted points or prediction time. The total prediction time is 1.0 s; that is, the total of predicted points is 600.

When Q equals 73 ml/min, the correlation coefficients are about 1.0 in the range of the prediction time interval. This indicates that long time predictions are reached with a high accuracy and the bubbling is nearly periodic. The reason why we use the word 'nearly' is that we note some decrease of the values of correlation coefficients when the prediction time is long. When Q varies from 192 to 238 ml/min, the correlation coefficients begin to decrease after N_p exceeds the value of about 250. The correlation coefficient fluctuations of the first few points are seen from Fig. 4.7, but this can't change the overall tendency. When Q increases further, the similar variation tendency can be obtained. For example, when Q is 988 ml/min, the correlation coefficients drop sharply after N_p exceeds the value of about 20. When Q is 2465 ml/min, the decreasing tendency of the correlation coefficient is irregular after N_p is larger than 20.

Figure 4.8 shows the correlation coefficients versus gas flow rates at a given number of predicted points, N_p. At low Q, the correlation coefficients are about 1.0 for all given points. When Q increasing, a slight drop of correlation coefficient is observed for small N_p and a sharp drop is seen for large N_p. When Q is very high, the correlation coefficients are about 1.0 only for $N_p = 1$, and those for other cases drop deeply.

From Figs. 4.6, 4.7 and 4.8, we can find the limited predictability of this bubbling system. Theoretically, limited predictability is a typical characteristic of the chaos motion for the noise-free signals. But for experimental time series, noise is unavoidable and in this case it may also be a typical characteristic of noisy determinism. Fortunately, the measures of noise reducing are taken and the other analyses are also carried out. Hence, the relationship between the limited

Fig. 4.8 Correlation coefficients between predicted and experimental values versus the gas flow rates [1, 2]

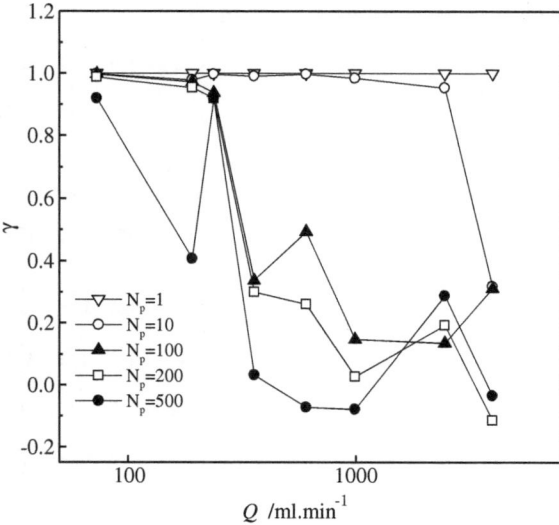

predictability and the chaos motion is closer. Limited predictability provides a strongest indicator of chaotic bubbling for experimental time series data.

It is noted that the jetting or a random bubbling process classified by this work when Q is very high, is different from the pure random process. This is due to the three features of the bubbling process. One feature is the long-term persistence or correlation appeared in the pressure signal. Another is the fine structure of the attractor restructured by the pressure signal. The last is the limited predictability of the pressure signal. These features are the characteristics of chaotic motion and not the features of the pure random motion. For the pure random motion, the accurate forecasting in time trajectory is very difficult.

These results of local non-linear short-term prediction can help us get direct evidence of chaotic bubbling and find the boundary between the chaotic bubbling and pure random process. Of course, we should keep in mind that the prediction length depends in some degree on certain parameters such as the number of data points M, the time delay τ, the attractor dimension D_2, etc. However, the local non-linear short-term prediction is undoubtedly a relative robust indicator of chaos identification.

4.2 Multi-Orifice Bubbling Behavior

4.2.1 Traditional Analyses

It is well known that the gas–liquid flow in a bubble column can be divided into two flow regimes with increasing superficial gas velocity: homogeneous bubbling flow regime and heterogeneous churn flow regime [6]. When the superficial gas

Fig. 4.9 **a** Homogeneous and **b** heterogeneous gas-liquid bubbling flow regimes

(a) 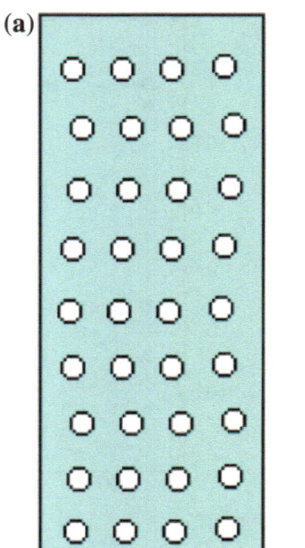 (b)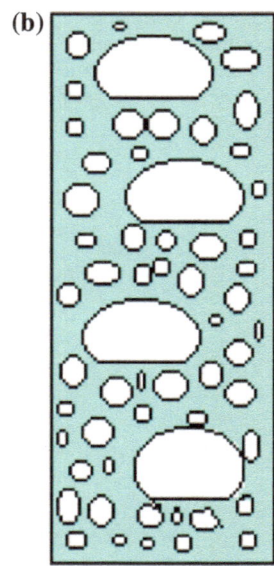

velocity is low, the flow is in the homogeneous bubbling flow regime. In this regime, the small and well-distributed bubbles rise through the bubble column in a sine trajectory at approximately the same speed. The size of the bubbles is nearly the same. There is no coalescence or breakup of the bubbles and the interaction between bubbles is weak. At certain superficial gas velocity, a sharp transition in the flow characteristics occurs and the flow goes into a new regime: the heterogeneous churn flow regime. In this regime, the bubbles begin to coalesce, forming fewer larger bubbles, and some smaller bubbles due to bubble break-up. The size of the bubbles is no longer uniform. Vortices of bubble swarms appear and rise in a more complex way due to intense interactions between bubbles. The space distribution of bubbles is very uneven. Visual observation in the gas–liquid bubble column confirms the above phenomena, as shown in Fig. 4.9. The superficial gas velocity for the regime transition between homogeneous bubbling flow regime and heterogeneous churn flow regime is about 4.965 cm/s, which is in the reasonable range of the transition velocity [7]. Some results obtained by using the traditional analytical techniques are shown in Figs. 4.10 and 4.11.

Figure 4.10 shows the time series of pressure fluctuations and corresponding power spectra at different superficial gas velocities. It can be seen from Fig. 4.10 that the time series and power spectra are very similar to those found in the literature [6]. The observed low-frequency effects in power spectra (as shown in Fig. 4.10b$_1$) in homogeneous bubbling flow regime are due to up- and down-streams of liquid. The weakening of the low-frequency effect is considered to be the precursor of flow regime transition.

Figure 4.11 shows the gas holdup and standard deviation of the pressure fluctuation signal. Little useful information for the flow could be derived from the time domain analysis and statistic analysis except for amplitude of the pressure

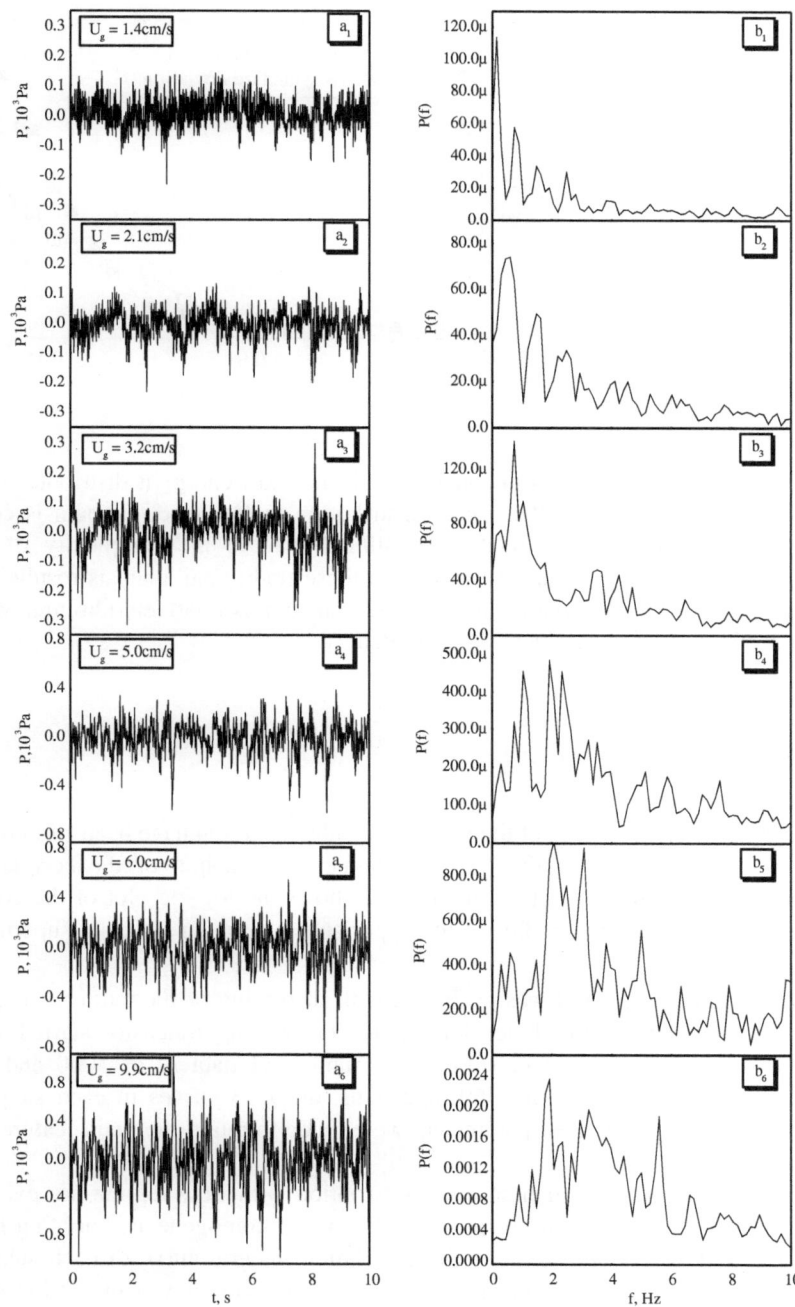

Fig. 4.10 Time series of pressure fluctuations and power spectrum at different superficial gas velocities in gas–liquid bubble columns [1, 8]. a_1–a_3, b_1–b_3 homogeneous bubbling flow regime; a_4–a_6, b_4–b_6 heterogeneous churn flow regime

Fig. 4.11 Gas holdup and
standard deviation versus
superficial gas velocity in
gas–liquid bubble columns
[1, 8]

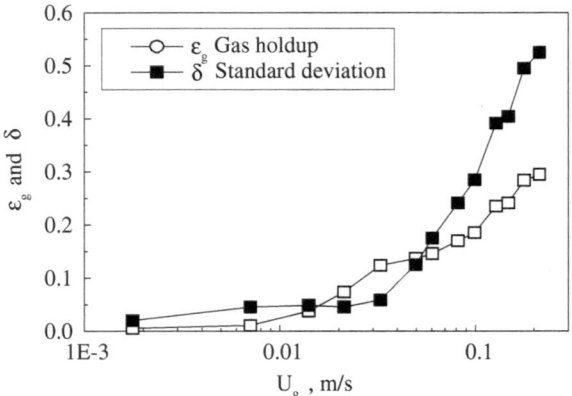

fluctuations. The power spectrum analysis indicates how energy is distributed over
the frequencies. However, the energy distribution in frequency domain is complicated and physical interpretation is difficult due to the complexity of the
physical source of the signal. According to these traditional analysis results, the
identification of the flow regime and regime transition is relatively obscure. More
accurate flow regime classification is needed.

4.2.2 Chaos Analyses

Chaos characteristic studies of the gas–liquid bubble column have been carried out
[9, 10]. In the present study, the multi-value phenomenon of the correlation
dimension is analyzed in detail. Figure 4.12 shows the log-log plot of the correlation integral $C_{(r)}$ versus radius of the hyper-sphere r, under different superficial
gas velocities, U_g.

The correlation integral was evaluated with increasing of the radius of hyper-
sphere at given embedding dimensions m. The embedding dimension started from
3 to 16 with time delay τ of 0.11 s, the number of data M of 5000 and the
restructured vector size N of 4500, and, thus, drew 13 curves in each subplot.
These choices of the related parameters were made to retain the main features of
the attractor.

The multi-value phenomenon and its variation with flow regime are evident
from the structure of the curves in Fig. 4.12. In the homogeneous bubbling flow
regime, as shown in Fig. 4.12a–c, the correlation integral curve of each subplot
exhibits basically only one linear section. The slope saturates at one value with
increasing embedding dimensions at any given superficial gas velocity, and, thus,
only one value of the correlation dimension can be obtained (typically $D_2 \approx 6.1$).
One value of the correlation dimension indicates that the structure of the attractor
in the phase space is well distributed in homogeneous bubbling flow regime

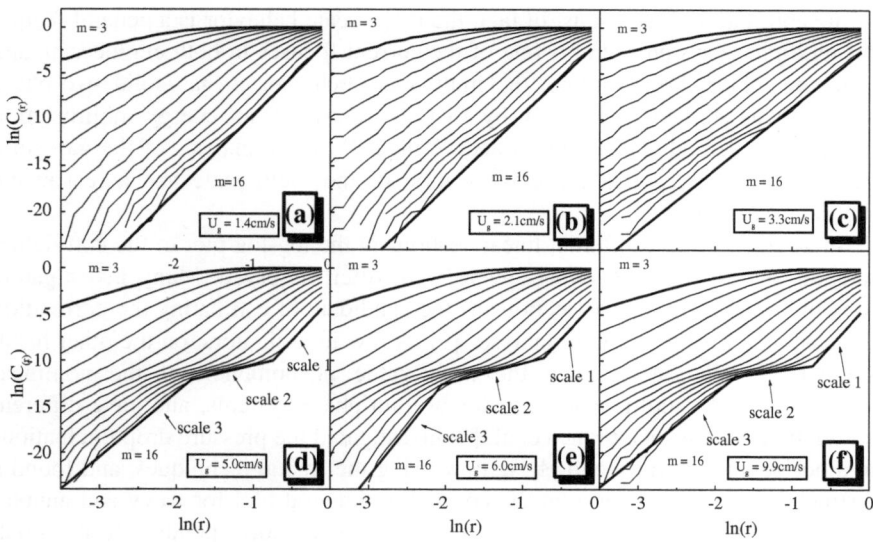

Fig. 4.12 Correlation integral versus radius of hyper-sphere at different flow regimes in bubble columns [1, 8]. **a–c** Homogeneous bubbling flow regime; **d–f** heterogeneous churn flow

because the correlation integral can be taken as the probability density according to its definition. This corresponds to the flow characteristics already mentioned of the homogeneous bubbling flow regime in bubble columns in real physical space.

When the flow changes from homogeneous bubbling flow regime to heterogeneous churn flow regime, more than one linear sections with nonzero slopes are obtained in each subplot of the correlation integral, as shown in Fig. 4.12d–f. In the present work, there are basically three longer linear sections, meaning that three values of the correlation dimension can be obtained at a specified superficial gas velocity. The appearance of three-slope curve implies that the structure of the attractor in the phase space is no longer well distributed in the heterogeneous churn flow regime. A hole or void arises in the restructured attractor. This corresponds to the flow characteristics of heterogeneous churn flow regime in bubble columns. It is noted that the curve changed its slope only when the embedding dimension is large enough ($m \geq 7$). The reason is that the attractor structure can be unfolded and, thus, the multi-dimension properties of the attractor that we wish to extract from the signals can be unraveled only when the embedding dimension is sufficient large.

The appearance of multi-value phenomenon of the correlation dimension is considered to be due to the presence of the multi-scale behavior in gas–liquid bubble columns. As already mentioned, the value of the correlation dimension represents the minimum number of the freedom variables, which are used to describe the system dynamics. Under some operating conditions, the gas–liquid flow manifests its behavior at several different scales [11]. It is not surprising that the dynamic behaviors are different for distinct scales, and, hence different values of the correlation dimensions. In fact, the multi-scale behavior is a general feature for multiphase systems, and is one of the current focuses in both engineering and academic fields. For example, the multi-scale behavior in gas–solid two-phase systems was studied extensively by Li [12], for which an energy minimization multi-scale method was proposed and verified. Wavelet analysis of the gas–solid flow further showed that the classification of the multi-scale had a reasonable physical basis [13].

In general, several different linear sections in the log-log plot of the correlation integral versus radius of hyper-sphere were often observed. Many investigators took only one of the linear sections as the candidate of estimating the correlation dimension according to their analyses [9]. A few investigators, on the other hand, studied the relation between the multi-value phenomenon of the nonlinear invariant and the multi-scale behavior in multiphase systems, and gave different correlation dimensions. Franca et al. [4] investigated the pressure drop fluctuations in a horizontal air/water two-phase flow using the fractal techniques, and found a distinct double-sloped curve in the correlation integral plot for wavy and annular flow. The authors speculated that large-scale signals (corresponding to the small correlation dimension $D_2 \approx 1.0$) came from surface waves while small-scale signals (corresponding to the large correlation dimension $D_2 \approx 6.1$) originated from rolling waves. Karamavruc and Clark [14] analyzed the local differential pressure time series measured from the gas–solid slugging bed using the deterministic chaos theory. For the signals recorded at low fluidization rates, the authors found that the correlation integral curve exhibits two different linear parts, and got two correlation dimensions, with $D_2 \approx 1.4$ for the highly periodic motion of slugs and $D_2 \approx 4.4$ for the irregular motion of particles. Bai et al. [15, 16] studied the chaotic behavior of gas–solid fluidized beds based on the signals of absolute and differential pressure and voidage. They also found two distinct linear sections of nonzero slope in the correlation integral curve and identified two correlation dimensions (called bi-fractal structure). They conjectured that the larger one ($D_2 \approx 7.5$), corresponding to small-scale fluctuations, represents particle motion and local turbulence, while the lower one ($D_2 \approx 6.5$), corresponding to large-scale fluctuations, represents bubble motion or other large-scale behavior. These studies indicated that the multi-value phenomenon of chaos characteristics have close relation with the multi-scale behavior and need further analysis. At the same time, the information hidden in the pressure signals is complicated and rich. It is well known that the pressure signals mainly reflect the global hydrodynamics due to the macroscopic flow and/or large bubbles. However, the local and microcosmic flow behavior can also be contained [4, 10, 14–16] and be explored through the chaos analysis.

For the gas–liquid bubble columns, the typical plot of the correlation integral versus radius of hyper-sphere in heterogeneous churn flow regime is shown in Fig. 4.12e. From left to right along the abscissa, one can see that, beyond the noisy region (in the left with a sharp slope and a short range of radius of hyper-sphere), the slope of the first linear section corresponding to a much smaller radius of hyper-sphere is much larger (typically, $D_2 \approx 10.6$). This linear section is mostly associated with the irregular motion of the small bubbles or drops, which are called micro-scale in this work. In heterogeneous churn flow regime, the formation of the micro-scale of small bubbles or drops results from the processes of intense coalescence and break-up of bubbles. The behavior of the micro-scale is characterized by relatively high frequency, large dimension and irregularity. Obviously, much more freedom variables are needed to describe the dynamics of micro-scale, and, thus, much larger value of the correlation dimension is expected.

In the middle size of the radius of hyper-sphere, a much smaller slope of the linear section was found. This section is considered to be associated with the regular motion of larger bubbles or bubble swarms, which are called meso-scale. The formation of the larger bubbles could arise from the coalescence of the smaller bubbles in the heterogeneous churn flow regime. The emergence of the meso-scale of larger bubbles is the deterministic component of the system signal. Its motion is featured by low frequency, small dimension and regularity. In order to describe the dynamics of the meso-scale, fewer freedom variables are enough, and, thus, much smaller value of the correlation dimension is expected. The calculated value of the correlation dimension of the meso-scale is about 1.51, which is close to the total dimension of periodic motion.

When the radius of hyper-sphere is considerably larger, the slope of this linear section is also much large (typically, $D_2 \approx 9.1$) but is relatively stable with increasing of superficial gas velocity as shown in Fig. 4.12d–f. The stable characteristic of the slope makes its physical source different from that of the first linear section at a much smaller radius of hyper-sphere. Because the slope of the first section changes (first increases and then reduces) gradually with increasing superficial gas velocity due to the intensive irregular motion of bubbles. This section is probably associated with the effects of the system boundary such as the column diameter on the flow, which is called macro-scale. Its actual physical meaning however needs further studies.

An important footnote to the above analyses is that the values of correlation dimensions estimated according to Fig. 4.12 are some large, however, they are still in a reasonable range (from 3 to 15). Although the measures of reducing noise have been taken, the pressure signals may be not very "clean" and further efforts are to be made. Of course, the manifold structure of the dynamics itself shouldn't be altered when the measures of reducing noise are to be taken.

4.2.3 Flow Regime Identification Method

Before and after the regime transition, the structures of the correlation integral plots are clear different. In the same flow regime, however, they are very similar. The structure corresponds to the flow regime one to one. We can reasonably conclude that the structure and the variation of the plot of correlation integral versus radius of hyper-sphere can be used to identify the flow regime and regime transition. This is tentatively called the method of *correlation integral analysis*.

Table 4.1 Relationships between flow regimes, superficial gas velocity and chaotic parameter values [1, 17]

U_g (cm/s)	Flow regime	LLE (bits/s)	K (bits/s)	D_2
0.00–2.13	Homogeneous bubble regime	0.00–0.10	14.0–15.0	4.00–5.23
2.13–4.97	Transition flow I	0.10–0.22	15.0–22.0	5.23–6.41
4.97–9.93	Vortical-spiral flow regime	0.22–0.32	22.0–22.5	6.41–6.51
9.93–12.78	Transition flow II	0.32–0.51	22.5–24.0	6.51–6.90
12.78–21.29	Slug flow regime	0.51–1.11	24.0–25.0	6.90–7.00

Fig. 4.13 LLE as a function of superficial liquid velocity [1, 17]

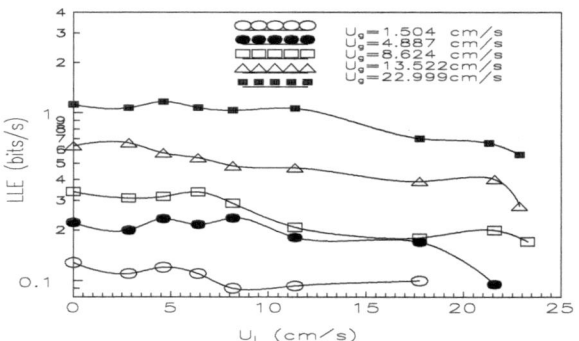

The structure of the correlation integral plot largely characterizes the intrinsic multi-scale behavior of the systems and should not be significantly affected by moderate changes of some other factors. On the other hand, the flow regime is determined according to the distinguished structure variation of the correlation integral plot in the present method and is not on the changes in the value of certain relevant invariant. This can eliminate some problems such as the scatter of the values and the disagreement on the varying tendency of the relevant invariant. Hence, even though the present identification method of flow regime for gas–liquid bubble columns is still in its early stage, and belongs to a qualitative one, at most the semi-quantitative one, it may be more robust, and, thus, more objective and more secure than the old ones. It can also be carried out rapidly like the power spectrum method and is compatible with on-line control of industrial reactors.

More experiments were also carried out in the gas–liquid bubble column with normal gas distributor under wider operation conditions and the chaotic hydrodynamics of gas–liquid two-phase flow was studied by the nonlinear analysis of time series of pressure fluctuations [1, 17]. The flow regime and its transition in two-phase bubble columns are more complex or rich and can be characterized by chaotic parameters such as the largest Lyapunov exponent, Kolmogorov entropy and correlation dimension. The main results are shown in Table 4.1.

The chaos properties of gas–liquid two-phase bubble columns are affected remarkably by the operation variables. When increasing superficial gas velocity, chaotic characteristic parameters increase, and they decrease when the superficial liquid velocity increases, as shown in Fig. 4.13.

Fig. 4.14 LLE as a function
of dimensionless axial
distance [1, 17]

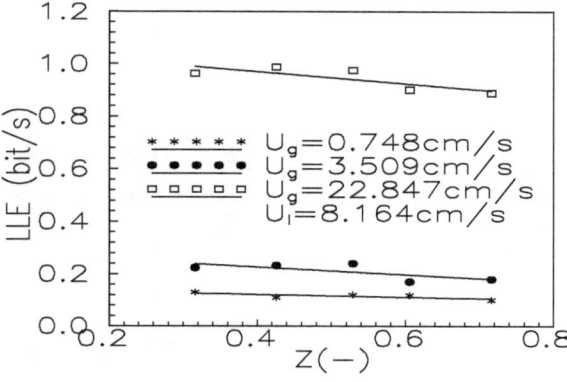

Fig. 4.15 LLE as a function
of dimensionless radial
distance [1, 17]

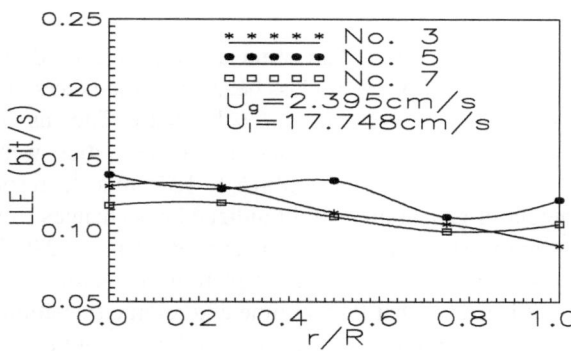

Table 4.2 Vertical distance of manometer taps above gas distributor [1, 17]

No.	1	2	3	4	5	6	7	8	9
Distance (m)	0.17	0.33	0.49	0.66	0.82	0.94	1.11	1.26	1.41

However, the chaotic characteristic parameters change relatively little in the
axial and radial directions of the bubble columns. Typical results are shown in
Figs. 4.14 and 4.15. In Fig. 4.15, No. 3, 5 and 7 mean different axial positions, as
shown in Table 4.2. The chaos analysis can be a quantitative technique for
identification of flow regime and regime transition in gas–liquid two-phase bubble
columns.

4.3   Gas–Liquid–Solid Fluidized Beds

4.3.1   Ordinary Gas–Liquid–Solid Fluidized Beds

The chaotic hydrodynamic behavior of flow in a gas–liquid–solid three-phase
fluidized bed was also studied by deterministic chaos analysis of time series data of
pressure fluctuations [1, 18]. The three-phase fluidized bed was of 0.07 m ID with

Fig. 4.16 Flow regimes for
the gas–liquid–solid fluidized
bed based on chaotic analyses
(*solid line*) and linear
analyses (*dashed line*) [1, 18].
CBFR coalesced bubble flow
regime, *DBFR* dispersed
bubble flow regime, *SBFR*
slug bubble flow regime, *TFR*
transition flow regime

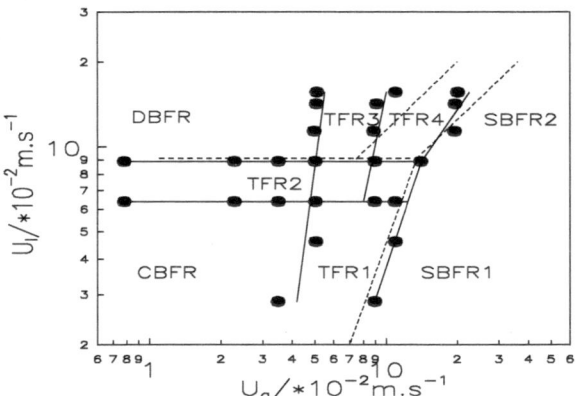

a height of 1.55 m. Tap water and air were used as liquid phase and gas phase, respectively. Spherical glass beads with diameters of 0.001–0.004 m and a density of 2500 kg m^{-3} were used as solid phase. The gas–liquid flow was co-current and upward. Superficial gas and liquid velocities were varied over the range of 0.757–26 × 10^{-2} and 0.0–16 × 10^{-2} m s^{-1}, respectively. The effective bed height of gas–liquid–solid fluidized beds changes from 0.8 to 1.3 m. Pressure data were measured with a transducer of IC-Sensor 1220-2psid. The sensor of the probe was placed 0.49 m above the plate horizontally.

It is shown that flow regime and regime transition in three-phase fluidized beds can be quantitatively characterized by nonlinear characteristic parameter—the largest Lyapunov exponent. A new flow regime map of three-phase fluidized beds that is more detailed than the traditional ones was drawn based on the largest Lyapunov exponent values, as shown in Fig. 4.16.

It can be seen from Fig. 4.16 that there is always a transition flow regime (TFR) between typical coalesced bubble flow regime (CBFR), dispersed bubble flow regime (DBFR) and slug bubble flow regime (SBFR) in the three-phase fluidized beds. But traditional linear analyses of flow regimes can find these transition flow regimes. Meanwhile, the slug bubble flow regime (SBFR) in a three-phase fluidized bed can be divided into two different sub-regimes according to the chaotic characteristic parameters. The hydrodynamics of one sub-regime (SBFR1) is controlled by the solid phase behavior and the hydrodynamics of another one .(SBFR2) is controlled by the liquid phase behavior. The relationship between LLE and superfical liquid velocity when flow regime changing from SBFR1 to SBFR2 is shown in Fig. 4.17.

Similarly, two transition flow regimes were identified when the dispersed bubble flow regime transits to the slug bubble flow regime, TFR3 and TFR4. The relationship between LLE and superfical gas velocity when flow regime changing from DBFR to SBFR2 is shown in Fig. 4.18. When $U_g < 4.975 \times 10^{-2}$ m s^{-1}, typical DBFR exists, and LLE ranges from 0.10 to 0.13 bit s^{-1}. When $U_g \geq 4.975 \times 10^{-2}$ m s^{-1}, DBFR begins to change and LLE > 0.20 bit s^{-1}. This unstable flow is called TFR3. When $U_g = 8.839 \times 10^{-2}$ m s^{-1}, the transition

Fig. 4.17 LLE versus superficial liquid velocity when flow regime changing from SBFR1 to SBFR2 [1, 18]

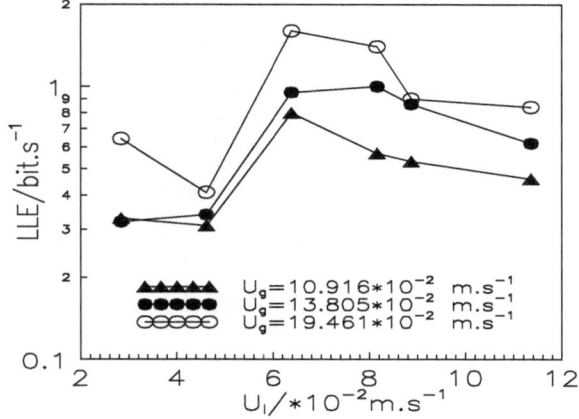

Fig. 4.18 LLE versus superficial gas velocity when flow regime changing from DBFR to SBFR2. $U_l = 11.359 \times 10^{-2}$ m s^{-1}, $d_p = 1.0$–1.25×10^{-3} m [1, 18]

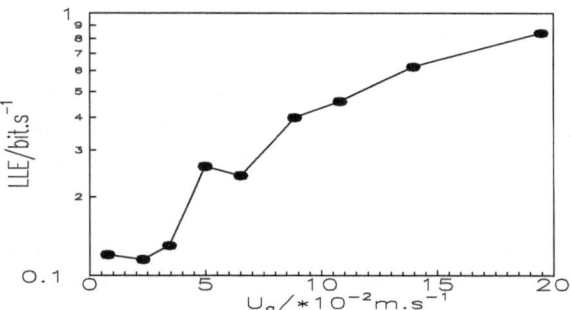

ends. Large slug bubbles can be found and LLE begins larger than 0.40 bit s^{-1}. Hence, the superfical gas velocity in TFR3 ranges from 4.975 to 8.839 $\times$ 10^{-2} m s^{-1} and LLE changes from 0.20 to 0.40 bit s^{-1}.

When $U_g > 8.839 \times 10^{-2}$ m s^{-1} and LLE > 0.40 bit s^{-1}, a different transition regime TFR4 appears. The superfical gas velocity in TFR4 ranges from 8.839 to 19.461 $\times$ 10^{-2} m s^{-1}, and LLE ranges from 0.40 to 0.86 bit s^{-1}. When further increasing superfical gas, the motion character of the large bubble changes again. Hence, this transition regime is called TFR4. When $U_g > 19.461$ m s^{-1}, LLE > 0.86 bit s^{-1} and the large bubble becomes pure slug one and moves up and down like the motion feature of a spring. And the flow regime is SBFR2.

4.3.2 Magnetized Gas–Liquid–Solid Fluidized Beds

The chaos characteristics of a gas–liquid–solid fluidized bed under a uniform magnetized filed were also studied by the deterministic chaos analysis of time series data of pressure fluctuations [1, 19]. The nonlinear characteristic parameters including the largest Lyapunov exponent and Kolmogorov entropy are all positive

Fig. 4.19 Effect of magnetic field on chaos characteristic parameters LLE of the gas–liquid–solid magnetized fluidized bed [1, 19]

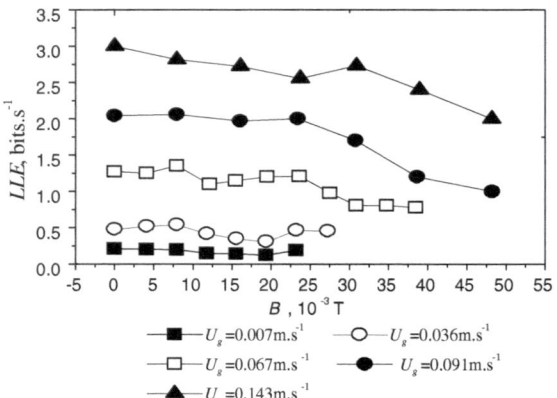

and they decrease with the increase of the magnetic field strength at given superficial gas velocity and liquid velocity. Typical results are shown in Fig. 4.19.

These results indicated that the system is chaotic and the magnetic field can make the three-phase flow more regular. The particulate flow regime, chain flow regime and magnetically aggregated flow regime in the three-phase magnetized fluidized bed can be characterized basically by these chaotic characteristic parameters.

The nonlinear analysis on the hydrodynamics in the self-aspirated reversed flow jet loop reactor was also carried out [1, 20]. In addition, the application energy-minimization multi-scale method to the gas–liquid–solid fluidized beds is also a related nonlinear topic [21].

References

1. M.Y. Liu, Studies on the chaos hydrodynamic characteristics of multiphase reactors. Dissertation, Tianjin University, 1998
2. M.Y. Liu, Z.D. Hu, Studies on the hydrodynamics of chaotic bubbling in a gas-liquid bubble column with a single orifice. Chem. Eng. Technol. **27**, 537–547 (2004)
3. S.Y. Xie, B.H. Tan, Bubble formation at multiple orifices-bubbling synchronicity and frequency. Chem. Eng. Sci. **58**, 4639–4647 (2003)
4. F. Franca, M. Acikgoz, R.T. Lahey, A. Clausse, The use of fractal techniques for flow regime identification. Int. J. Multiphase Flow **17**, 545–552 (1991)
5. M.C. Ruzicka, J. Drahos, J. Zahradnik, N.H. Thomas, Intermittent transition from bubbling to jetting regime in gas-liquid two phase flows. Int. J. Multiphase Flow **23**, 671–682 (1997)
6. J. Drahos, F. Bradka, M. Puncochar, Fractal behaviour of pressure fluctuations in a bubble column. Chem. Eng. Sci. **47**, 4069–4075 (1992)
7. A. Sarrafi, M. Jamialahmadi, H. Muller-Steinhagen, J.M. Smith, Gas holdup in homogeneous and heterogeneous gas-liquid bubble column reactors. Can. J. Chem. Eng. **77**, 11–21 (1999)
8. M.Y. Liu, J.H. Li, Z.D. Hu, Multi-scale characteristics of chaos behavior in gas-liquid bubble columns. Chem. Eng. Commun. **191**, 1003–1016 (2004)

9. H.M. Letzel, J.C. Schouten, R. Krishna, C.M. van den Bleek, Characterization of regimes and regime transitions in bubble columns by chaos analysis of pressure signals. Chem. Eng. Sci. **52**, 4447–4459 (1997)
10. M. Cassanello, F. Larachi, M. Marie, C. Guy, J. Chaouki, Experimental characterization of the solid phase chaotic hydrodynamics in three-phase fluidization. Ind. Eng. Chem. Res. **34**, 2971–2980 (1995)
11. B.R. Bakshi, H. Zhong, P. Jiang, L.S. Fan, Analysis of flow in gas-liquid bubble columns using multi-resolution methods. Trans. IchemE Part A **73**, 608–614 (1995)
12. J.H. Li, Multi-scale modeling and method of energy-minimization in gas-solid two-phase flow. Dissertation, Institute of Chemical & Metallurgy, Chinese Academy of Sciences, 1987
13. J.Q. Ren, Wavelet analysis of dynamic behavior in fluidized beds. Dissertation, Institute of Chemical & Metallurgy, Chinese Academy of Sciences, 1999
14. A.I. Karamavruc, N.N. Clark, Local differential pressure analysis in a slugging bed using deterministic chaos theory. Chem. Eng. Sci. **52**, 357–370 (1997)
15. D. Bai, E. Shibuya, N. Nakagawa, K. Kato, Fractal characteristics of gas-solids flow in a circulating fluidized bed. Powder Technol. **90**, 205–212 (1997)
16. D. Bai, A.S. Issangya, J.R. Grace, Characteristics of gas-fluidized beds in different flow regimes. Ind. Eng. Chem. Res. **38**, 803–811 (1999)
17. M.Y. Liu, Z.D. Hu, Chaos analysis of flow regime and regime transition in gas-liquid two-phase bubble columns. Eng. Chem. Metall. **21**, 37–43 (2000)
18. M.Y. Liu, Z.D. Hu, Chaos analyses of flow regime and regime transition in gas-liquid-solid three-phase fluidized beds. Chem. React. Eng. Technol. **16**(4), 363–368 (2000)
19. M.Y. Liu, J.Y. Wu, Z.D. Hu, Chaos characteristics in a gas-liquid-solid three-phase magnetized fluidized bed. J. Chem. Eng. Chin. Univ. **13**, 476–480 (1999)
20. M.Y. Liu, J.P. Wen, X.Y. Qin, Z.D. Hu, Local chaos characteristics in a self-aspirated reversed flow jet loop reactor. Trans. Tianjin Univ. **14**, 56–59 (1998)
21. M.Y. Liu, J.H. Li, M. Kawauk, Application of the energy-minimization multi-scale method to gas-liquid-solid fluidized beds. Chem. Eng. Sci. **56**, 6805–6811 (2001)

Chapter 5
Concluding Remarks

Investigations of the non-linear bubbling hydrodynamics in a gas–liquid bubble column with a single orifice and a multi-orifice gas distributor, in a gas–liquid–solid fluidized bed and a gas–liquid–solid magnetized fluidized bed were reported in this chapter. The main concluding remarks are as follows.

For the single-orifice gas–liquid bubble column, bubbling process exhibited a deterministic chaos in a certain range of gas flow rates, and with the increase of the gas flow rate, the bubbling undergoes in turn four different stages: period bubbling, primary chaotic bubbling, advanced chaotic bubbling and jetting. However, no clear period doubling sequence leading to chaotic behavior was observed. The non-linear invariants such as K and D_2 together with the plot of the correlation integral may be developed as potential effective quantitative tools for flow regime identification of a bubbling process. The sudden increase of local nonlinear short-term forecasting error of pressure signal with time or number of predicted points (i.e. the sharp loss of the ability to predict the pressure signal successfully with the non-linear forecasting method) was an evidence of the presence of the chaotic bubbling, and also an effective way of distinguishing chaos from pure random.

For the multi-orifice gas–liquid bubble columns, the flow regimes and their transitions could be characterized by the chaotic invariants and the structure of the correlation integral plot. Similarity in structure of the correlation integral plot showed similar flow regime. Only one linear section in the correlation integral plot meant that the flow was in the homogeneous bubbling flow regime, and the correlation integral plot exhibiting a multiplicity of linear sections indicated that the flow was in the heterogeneous churn flow regime. The appearance of multi-value phenomenon of the correlation dimension for heterogeneous churn flow regime resulted from the formation of multi-scale behavior in gas–liquid bubble columns.

For the gas–liquid–solid fluidized bed, the nonlinear analysis is also a powerful tool to study their hydrodynamics. The flow regimes and their transitions can also be identified effectively by the chaotic invariants and plots and some new transition flow regimes were found with the help of the chaotic analyses.

For the gas–liquid–solid magnetized fluidized bed, the chaotic invariants reduce with the increase of the magnetic field strength at given superficial gas velocity

M. Liu and Z. Hu, *Nonlinear Analysis and Prediction of Time Series in Multiphase Reactors*, SpringerBriefs on Multiphase Flow, DOI: 10.1007/978-3-319-04193-3_5, © The Author(s) 2014

and liquid velocity, which indicate that magnetic field can make the three-phase flow more regular.

Of course, the flow regime identification methods built in this chapter have their limitations of applications because the sizes of the reactors didn't vary and the experimental conditions are also not very wide. Much work is needed to find more general, objective, accurate and robust methods for flow regime identification.

It is worth mentioning that this chapter mainly focused on the issues of chaotic analysis and prediction of time series of physical parameters measured from several multiphase reactors. Chaos analysis results show that these multiphase systems are nonlinear and not linear ones. Successful nonlinear prediction on the evolution behavior of physical parameter itself revealed the chaotic bubbling dynamic mechanism in the single-orifice bubbling system. Further work will focus on exploring the physical mechanisms hidden in these complex nonlinear multiphase dynamic systems and modeling or simulating dynamic behaviors successfully.